Solid State Fundamentals
for Electricians

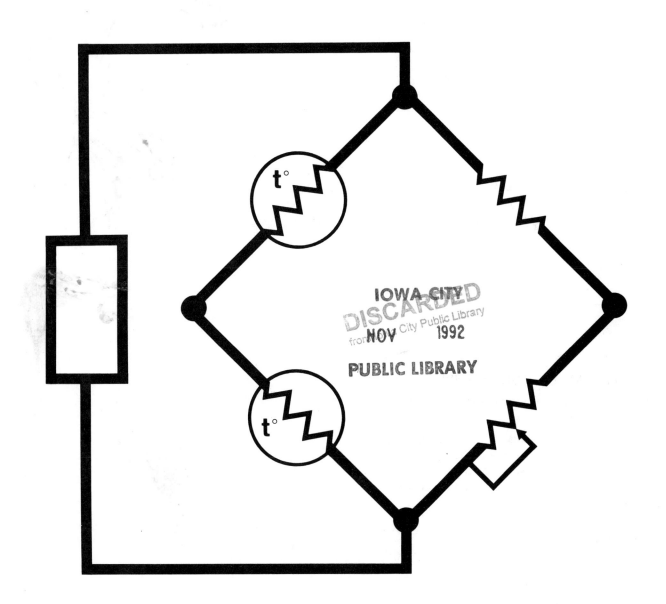

AMERICAN TECHNICAL PUBLISHERS, INC.
HOMEWOOD, ILLINOIS 60430

Gary Rockis

Preface

SOLID STATE FUNDAMENTALS FOR ELECTRICIANS presents a comprehensive overview of solid state devices and systems, including fiber optics, integrated circuits, and light-activated components. The textbook is designed for electricians, students, and others who have some basic knowledge of electronics. Component and system construction, operation, installation, and service are emphasized. Various practical applications are presented throughout the textbook as they relate to temperature, light, speed, and pressure control. It is a practical textbook; however, theory and concepts are adequately covered to assure complete understanding of the topics presented.

The textbook contains 12 chapters with 485 illustrations, including schematic diagrams, characteristic curves, and charts to support the text. The graphic symbols used are based on standards from the American National Standards Institute (ANSI) and the Institute of Electrical and Electronics Engineers (IEEE), as well as industry norms—especially in the area of fiber optics and light-activated devices. Second color is used in the textbook to help the reader assimilate the material quickly by emphasizing important points, components or parts of components, and correlating the illustrations to the information presented in the text.

Key words are listed at the beginning of each chapter and are clearly defined and/or illustrated within the chapter. Review Questions are provided at the end of each chapter to test comprehension of the material presented. A thorough index is at the back of the textbook. *SOLID STATE FUNDAMENTALS FOR ELECTRICIANS* is not only a practical textbook for individuals in the trade or for students of electricity and electronics, but it is also a valuable source of reference.

The Publishers

ACKNOWLEDGEMENTS

Cover was designed by:
    Handelan Pedersen, Inc.

Illustrations were drawn by:
    AZ-TECH Design & Graphics, Inc.

Additional Art was provided courtesy of:
    Energy Concepts, Inc.
    Micro Switch, A Division of Honeywell
    National Safety Council

CONTENTS

1 SAFETY

Safety is important to everyone. Accidents result in damage to property, loss of time, increased cost, personal injury, or possibly a fatality. Electricians are aware of the dangers associated with electrical power and the potential dangers that can exist on a job or at a training facility. Never be complacent about safety. By exercising common sense and practicing safe work habits, accidents can be avoided.

Key Words

Carding	Class B Fire	Electrical shock
Cardiopulmonary resuscitation (CPR)	Class C Fire	Ground
Class A Fire	Class D Fire	

(National Safety Council)

HAND TOOL SAFETY

When selecting hand tools, choose the right one for that particular job. Make sure it is the correct size. Use good quality tools and use them for the job they were intended to perform. For example, it is often tempting to use a screwdriver for a chisel or a pair of pliers for a wrench—avoid such practices. The right tool will do the job faster and safer. Never force a tool or use a tool beyond its capacity. The price of the tool and the time required to replace it will prove far less costly than a serious accident.

Maintaining Hand Tools

Keep tools in good working condition. Periodic inspection of hand tools will help keep them in good condition. Always inspect a tool before using it. Never use a tool that is in poor or faulty condition. Tool handles should be free of cracks or splinters and should be fastened securely to their working part. Damaged tools are not only dangerous, but are also less productive than those in good working condition. When inspection shows a dangerous condition, repair or replace the hand tool immediately.

Storing Hand Tools

Keep hand tools in a safe place. Many accidents are caused by tools falling off ladders, shelves, or scaffolds that are being moved. Each tool should have a designated place in the tool box or electrician's pouch.

(National Safety Council)

Never carry tools in a pocket unless the pocket is designed for that purpose. Never place pencils behind the ear or under a hat or cap.

Keep cutting tools away from the edge of a work bench. Brushing against the tool may cause it to fall and result in an injury to a leg or a foot. When carrying edged or sharply pointed tools, carry them with the cutting edge or the point in a downward fashion away from the body.

POWER TOOL SAFETY

Power tools, like hand tools, must be properly maintained and stored. In addition, power tools should be carefully monitored and be under control any time they are turned ON.

Power Tool Safety Checklist

The following checklist is helpful in maintaining control of power tools:

1. Be thoroughly familiar with all of the manufacturer's safety recommendations and the operating features of the power tool before using it.

2. Be sure that all safety guards are in place and in working order.

3. Wear safety goggles or a face shield, and a respirator or dust mask when conditions require them.

4. Be sure that the material to be worked on is free of obstructions and properly secured.

5. Before connecting a power tool to a power source, be sure that it is turned OFF.

6. Keep your attention focused on the work. Never allow yourself to be distracted.

7. A change in power tool sound during operation usually indicates a problem—investigate immediately.

8. Power tools should be inspected and serviced by a qualified repair person at regular intervals. Follow the manufacturer's recommendations concerning all service and maintenance requirements.

9. Inspect electrical cords for fraying or other damage. Correct any potential problem immediately.

10. When work is completed, turn OFF the power. Wait until the power tool stops before leaving a stationary tool or before laying down a portable tool.

Grounding Power Tools

All power tools must be properly grounded, unless they are approved double-insulated tools. Power tools with a three-prong plug should be connected to a grounded outlet (receptacle). It is very dangerous to use an adapter to connect a three-prong plug into a two-hole conductor outlet unless a separate ground wire is connected to an approved *ground* (common return circuit). The ground ensures that any short will trip the circuit breaker or blow the fuse.

WARNING: An ungrounded power tool is a potential for a fatal accident.

Double-insulated tools have two-prong plugs and are properly identified as such by the manufacturer on the nameplate. The electrical parts of the motor in a double-insulated power tool are provided with extra insulation to prevent electrical shock. Therefore, the tool does not have to be grounded. Both the interior and exterior should be kept clean and free from grease or dirt that might conduct electricity. Follow the manufacturer's recommendations regarding service and maintenance.

CLOTHING AND PERSONAL SAFETY

The clothing worn at work is important for personal safety. Appropriate attire should be worn for each particular job site and work activity. The following points should be observed:

1. Wear thick-soled work shoes for protection against sharp objects such as nails. Wear shoes with safety toes if the job requires it. If shoes are subjected to oil and grease, make sure the soles are oil resistant.

2. Wear rubber boots in damp and wet locations.

3. Wear an approved safety helmet (hard hat) if the job requires it. Confine long hair or keep hair trimmed when working around machinery.

4. Never wear jewelry on the job. Gold and silver are excellent conductors of electricity.

(National Safety Council)

ELECTRICAL SAFETY

An electrician must understand and be able to apply the principles of electricity and safety. The time to prevent an accident is before it happens. Respect for electricity comes from understanding it and its potential hazards. Report any unsafe condition, unsafe equipment, or unsafe work practices to a supervisor as soon as possible.

Fuses

Before removing any fuse from a circuit, be sure the switch for the circuit is open or disconnected. When removing cartridge fuses, use an approved fuse puller and break contact on the hot side (line side) of the circuit first. When replacing cartridge fuses, install the fuse into the load side of the fuse clip first, and then into the line side.

Electrical Shock

Electrical shock occurs when a person comes in contact with two conductors of an electrical circuit or when the body becomes part of the electrical circuit. A severe electrical shock can cause the heart and lungs (cardiopulmonary system) to stop functioning. Also, severe burns may occur where the current enters and exits the body. Prevention is the best medicine for electrical shock. Respect all voltages, understand the circuitry, and follow safe work procedures. Do not take chances.

Electricians should know the correct procedure for rescuing an electrical shock victim. If possible, always turn OFF or disconnect the circuit before touching the victim. If this is not possible, use a nonconductive object, such as a stick of wood, to break the contact of the person from the circuit, then pull the victim away. If the circuit cannot be turned OFF, and the contact (victim and circuit) cannot be broken, wear rubber gloves and stand on a nonconductive dry material, then pull the victim free. Time can be a critical factor, but exercise common sense before taking action.

All electricians are encouraged to take a basic course in *cardiopulmonary resuscitation (CPR)* so they can come to the aid of a co-worker in emergency situations.

Out-of-Service Protection

Before servicing any electrical equipment, be absolutely certain that the power source is turned OFF and properly carded (out of service). *Carding* is the process of padlocking, in the OFF position, the power source, and indicating on an appropriate card the procedure that is taking place.

Whenever leaving the job site for any reason, or whenever the job cannot be completed in the same day, be sure that the power source is turned OFF. Check the power source before returning to work.

SAFETY COLOR CODES

The Occupational Safety and Health Administration (OSHA) has established the following colors to designate certain cautions and dangers:

Red: Fire protection equipment and apparatus.
Portable containers of flammable liquids.
Emergency stop buttons and switches.

Yellow: Caution and for marking physical hazards.
Waste containers for explosive or combustible materials.
Caution against starting, using, or moving equipment under repair.
Identification of the starting point or power source of machinery.

Orange: Dangerous parts of machines.
Safety starter buttons.
The exposed parts (edges) of pulleys, gears, rollers, cutting devices, and power jaws.

Purple: Radiation hazards.

Green: Safety.
Locations of first aid equipment (other than fire fighting equipment).

FIRE SAFETY

The chance of fire can be greatly reduced by practicing good housekeeping. Keep debris in a designated area away from the building. In any event, if a fire should start, alert all workers on the job and notify the fire department. Before the fire department arrives, a reasonable effort should be made to contain the fire. However, no action should be taken that could jeopardize anyone's safety.

Fire Extinguishers

The following classifications of fire extinguishers are available: A, A/B, B/C, A/B/C, and D. The letter designation of a particular fire extinguisher indicates the class(es) of fire(s) it can extinguish. The classes of fires are:

Class A Fire: Ordinary combustible materials such as wood, cloth, paper, rubber, and many plastics.

Class B Fire: Flammable liquids/gases, and greases. *Only dry chemical types of extinguishers are effective on pressurized flammable gases/liquids. For deep fat fryers, multipurpose A/B/C dry chemicals are NOT acceptable.*

Class C Fire: Energized electrical equipment. The electrical nonconductivity of the extinguishing media is important.

Class D Fire: Combustible metals such as magnesium, titanium, zirconium, sodium, and potassium.

(National Safety Council)

In-Plant Firefighting Training

It is highly recommended that a selected group of personnel (if not all personnel) be acquainted with all types and sizes of fire extinguishers in the plant or work area. Such training should include familiarization with the facility layout and location of fire extinguishers. Regular practice sessions should also be held.

Fire Extinguisher Maintenance

Like any other piece of equipment, fire extinguishers require routine maintenance. The following points should be observed:

1. Inspect extinguishers at least once a month. It is very common to find units that are missing, damaged, or discharged.

2. Never attempt to make repairs to fire extinguishers. This is the chief cause of dangerous shell ruptures.

3. Consider contracting for maintenance with a reputable fire extinguisher organization.

Chapter 1 - Review Questions

1. What safety precautions are applicable to all hand tools?
2. List five safety checks for power tools.
3. When may an adapter be safely used with a power tool that has a three-prong plug?
4. How is a cartridge fuse properly removed and replaced?
5. Explain the procedure for freeing an electrical shock victim.
6. What is carding?
7. According to OSHA's Safety Color Code, what does the color green designate?
8. What is a Class C fire?

2 PC BOARD CONSTRUCTION AND REPAIR

The demand for miniaturization of electronic equipment has led to many different construction techniques. One of the most popular techniques used is the *printed circuit assembly*, or *PC board*.

Key Words

Bar-Type Desoldering Tip	Dual In-Line Desoldering Tip	Oxidation
Board Extractors	Edge Card	Pads
Bridging	Edge Card Connector	PC Board
Bus	Eutectic	Printed Circuit Assembly
Coated Tip	Foils	Seizer
Cold Solder Joint	Foil Separation	Solder Sucker
Conformal Coating	Heat Sink	Tinning
Cup-Type Desoldering Tip	Holding Jig	Traces

PC BOARD CONSTRUCTION

A PC board is made of an insulating material such as fiberglass or phenolic with conducting paths secured to one or both sides of the board. PC boards are manufactured through a variety of processes. The purpose of all PC boards is to provide electrical paths of sufficient size to ensure a reliable electronic circuit.

Each part of the PC board has a name. See Figure 2-1. *Pads* are the small round conductors to which component leads are soldered. *Traces*, or *foils*, are used to interconnect two or more pads. A large trace or foil extending around the edge to provide conduction from several sources is called a *bus*.

An *edge card* is a PC board with multiple terminations (terminal contacts) on one end. Most edge

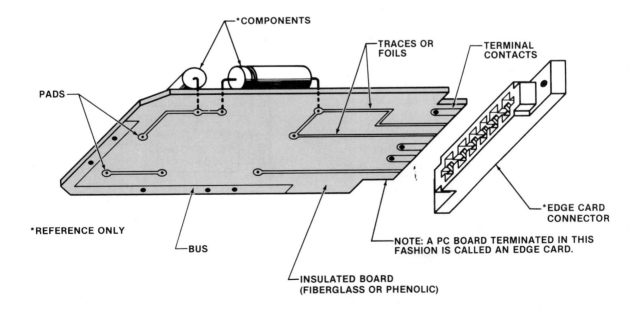

Figure 2-1. A PC board is an insulating material such as fiberglass or phenolic with conducting paths secured to one side or both sides of the board. Each part of the PC board has a name.

5

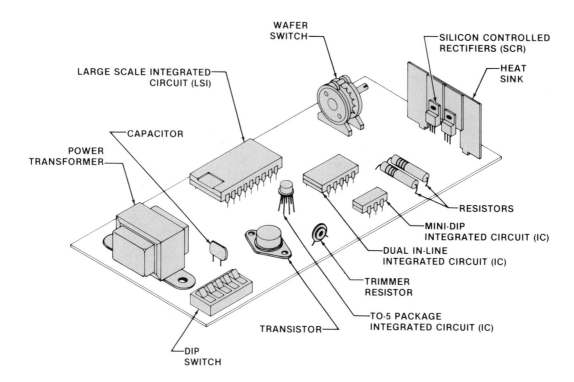

Figure 2-2. A wide variety of components may be mounted on a PC board. Component leads extend through the board and are connected to the pads, traces, and buses with solder.

cards have terminations made from the same material as the foils—namely copper. In some instances, the terminations are gold plated, allowing for the lowest possible contact resistance. An *edge card connector* allows the edge card to be connected to the system's circuitry with the least amount of hardware.

Components are usually mounted on one side of the PC board. See Figure 2-2. However, in some cases where space is a premium, components may be mounted on both sides of the PC board. Component leads extend through the insulated board and are connected to the pads, traces, and bus with solder. To expedite troubleshooting procedures, many PC boards have an appropriate marking next to the component. The marking identifies the component in relation to the schematic.

Locating Components on a PC Board

When troubleshooting equipment that includes PC boards, identification of PC board circuitry and its components should be relatively easy. Most manufacturers have developed methods of quick identification such as numbering and lettering the major components on the schematic. See Figure 2-3. Many manufacturers also provide a pictorial guide to locate components on the PC board by using an actual photograph of the PC board. See Figure 2-4.

Electrical Measurements on a PC Board

When taking resistance or continuity measurements on resistors, capacitors, and inductors, they are usually taken from the component side of the board rather than the foil side of the board. Voltage measurements are taken from either side of the board. Many PC

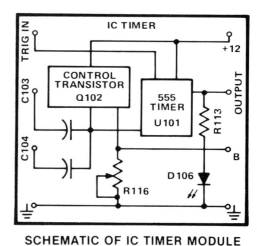

SCHEMATIC OF IC TIMER MODULE

Figure 2-3. Most manufacturers number and letter the major components on the schematic. *(Energy Concepts, Inc.)*

ACTUAL PC BOARD

Figure 2-4. Many manufacturers provide a pictorial guide to expedite location of components on the PC board. *(Energy Concepts, Inc.)*

boards are covered with a non-conductive protective coating (*conformal coating*). When taking measurements on this type of board, it may be necessary to use needle-point meter probes to penetrate the protective coating. See Figure 2-5. If necessary, scrape away part of the protective conformal coating. However, be sure that the conformal coating is repaired before putting the PC board back into service.

CAUTION: When taking measurements on live circuitry, be extremely careful not to short out components that could result in permanent damage to the component.

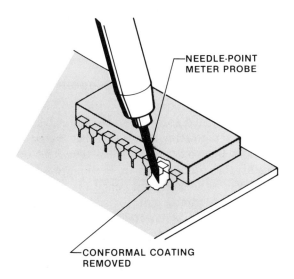

Figure 2-5. A needle-point meter probe may be needed to penetrate a PC board protected by a conformal coating, or to scrape away part of the coating.

Special probes are designed for use in close quarters to avoid shorting out components. See Figure 2-6. The special probes minimize slippage, thus reduce the likelihood of shorting out other components. In excessively crowded situations, a clip extender is attached to the component before the special probe is used. See Figure 2-6.

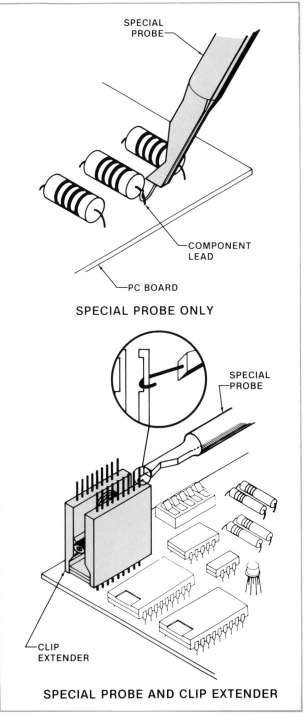

SPECIAL PROBE ONLY

SPECIAL PROBE AND CLIP EXTENDER

Figure 2-6. In close quarters, it may be necessary to use special probes to avoid shorting out surrounding components.

Removing PC Boards

CAUTION: When removing a PC board for repair, the circuit should be de-energized to eliminate possible damage to the circuit by surge currents. It also protects the service person from electrical shocks.

Edge cards are designed to fit snuggly into their edge card connectors. To remove them, it is often necessary to carefully work them back and forth in a rocking motion. See Figure 2-7. When the edge card is mounted in racks, a set of *board extractors* is used to dislodge it from the edge connector. See Figure 2-8.

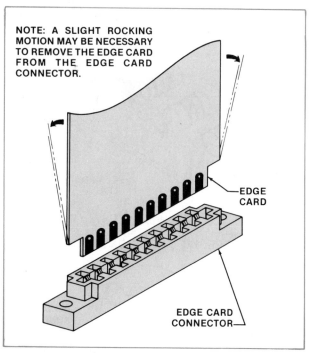

NOTE: A SLIGHT ROCKING MOTION MAY BE NECESSARY TO REMOVE THE EDGE CARD FROM THE EDGE CARD CONNECTOR.

EDGE CARD

EDGE CARD CONNECTOR

Figure 2-7. To remove an edge card, it is often necessary to carefully work it back and forth in a rocking motion.

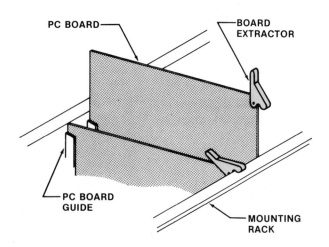

PC BOARD

BOARD EXTRACTOR

PC BOARD GUIDE

MOUNTING RACK

Figure 2-8. Board extractors are often used to remove the edge card.

PC BOARD SERVICE

PC boards incorporate several subminiaturized features not found in conventional equipment. Therefore, some specialized knowledge and tools are required for efficient servicing. A PC board must be held properly while repairs are made. Once a PC board is removed, it is good practice to use a *holding jig* to firmly secure it. See Figure 2-9. A holding jig also positions the board so that it is easy to work on, and it prevents the PC board from flexing as pressure is applied.

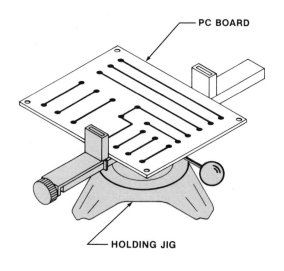

PC BOARD

HOLDING JIG

Figure 2-9. A PC board holding jig firmly secures a PC board for repairs.

Desoldering a PC Board Component

When removing a component, the PC board is generally positioned so that the component leads to be unsoldered are facing downward. See Figure 2-10. The tip of a hot soldering pencil is then held under and against the component lead. See Figure 2-11. The solder flows to the soldering pencil tip, but may be removed with a wiping cloth or sponge. See Figure 2-12. This procedure is repeated until sufficient solder is removed from the component lead to free that end of the component. When both leads are loose, lift the component from the PC board.

CAUTION: The part to be removed should NEVER be pried or forced loose. Any attempt to force a part loose may result in a broken or separated printed circuit panel.

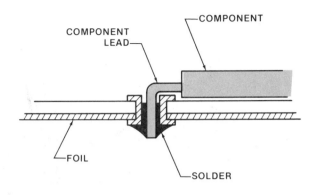

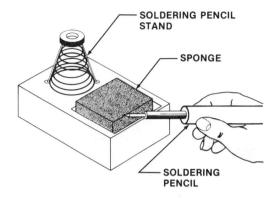

Figure 2-10. To remove a component, the PC board is positioned so that the component leads are facing downward.

Figure 2-12. A wet sponge or wet cloth may be used to remove solder from the soldering tip.

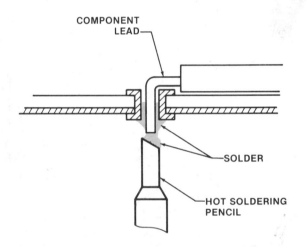

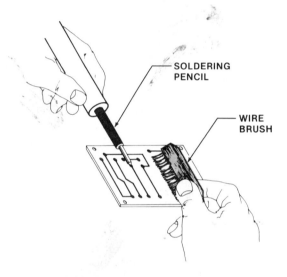

Figure 2-11. The tip of the soldering pencil is held against the component lead so that the solder will flow to the tip.

Figure 2-13. A component lead may be brushed free of melted solder with a small wire brush.

If any solder is left in the terminal hole after the component leads have been removed, lightly apply the soldering pencil tip to the hole to melt the solder. Clear the melted solder from the hole by using a reamer or small wire brush. See Figure 2-13.

There are also other techniques and tools used when desoldering PC board components: (1) A *solder sucker* used in conjunction with a soldering pencil sucks the solder away from the connection once the solder is melted. See Figure 2-14. (2) A specially woven strand of copper-coated wire is placed on the joint (connection) prior to applying a hot soldering pencil. See Figure 2-15. As the solder melts, capillary action draws all the solder out of the joint and into the woven strands of copper-coated wire. When finished, the used portion of woven strand is cut off and thrown away, making the spool ready for use again. (3) A combination soldering pencil and solder sucker is used when a large amount of solder is to be removed. See Figure 2-16.

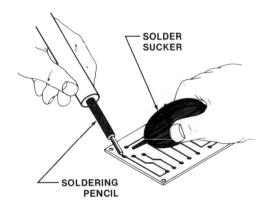

Figure 2-14. Solder is removed by using a solder sucker in conjunction with a soldering pencil.

Figure 2-15. Solder is removed by placing a specially woven strand of copper-coated wire on the joint prior to applying a hot soldering pencil.

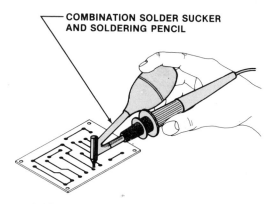

Figure 2-16. Solder is removed by a combination solder sucker and soldering pencil.

Isolating Open Circuits on a PC Board

Cracks and breaks on a PC board circuit can be difficult to see. It is often necessary to use a magnifying lens when examining the PC board for defects. See Figure 2-17.

If a PC board is suspected of having an open foil or trace (open circuit), an ohmmeter can also be used to check it. Take each section of the foil or trace and check for continuity. See Figure 2-18. If the resistance increases dramatically on a continuous trace, an open circuit is very likely. As the board is inspected, a certain amount of flexing may indicate a hairline crack that would not at first be noticeable.

If a break is located, any loose or separated foil should be cut away. See Figure 2-19. Removal of the loose foil to a secure point enables the repair to be made.

CAUTION: Care must be exercised so that hairline cracks are not created by too much flexing.

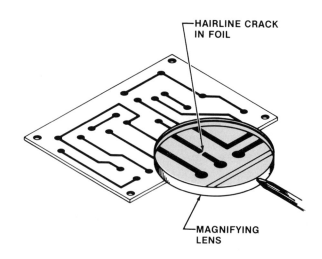

Figure 2-17. A magnifying lens can be used to find cracks or breaks in a PC board circuit.

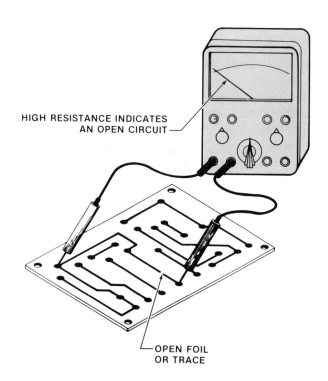

Figure 2-18. An ohmmeter can be used to check for open circuits on a PC board.

Repairing an Open Circuit on a PC Board

Once the foil has been cut back to a secure point, the foil section to be repaired should be cleaned with an appropriate solvent to remove any protective coatings. See Figure 2-20. With the protective coating removed, it is helpful to clean the surface of the foil. Steel wool could be used to remove any *oxidation* prior to soldering. See Figure 2-21.

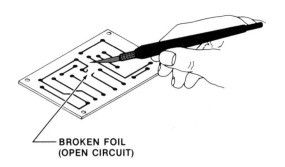

Figure 2-19. A broken foil should be cut back to a secure point on the PC board where the foil no longer lifts or separates.

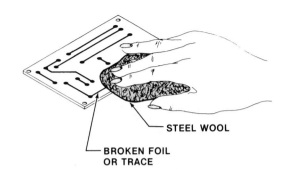

Figure 2-21. Steel wool should be used to remove any oxidation prior to soldering.

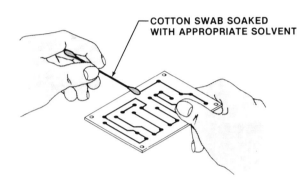

Figure 2-20. Protective coatings can be removed with a cotton swab and an appropriate solvent.

Repairing Small Breaks in Foil. In small breaks, a short piece of bare jumper wire soldered across the open foil is sufficient to bridge the gap. See Figure 2-22, Left. To repair this type of break:

1. *Tin* the short piece of jumper wire.
2. Place the jumper wire across the break.
3. Secure the jumper wire.
4. Solder, allowing the solder to flow along the length of the break so that the *tinned lead* becomes part of the circuitry.

CAUTION: Be sure to apply only enough heat to melt and flow the solder.

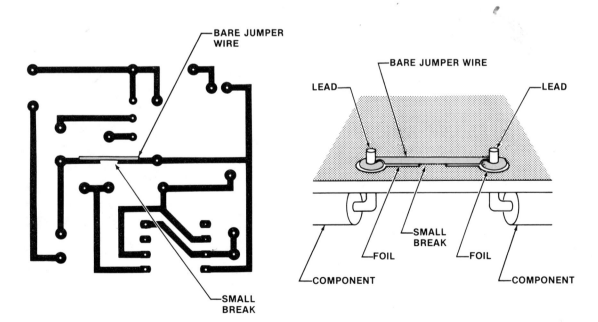

BARE JUMPER WIRE OVER BREAK   BARE JUMPER WIRE WRAPPED AROUND LEADS

Figure 2-22. For small breaks, solder a short piece of bare wire across the open foil, or solder a piece of bare jumper wire wrapped around component leads.

If components are in close proximity, the jumper wire may be wrapped around the leads to secure it before soldering. See Figure 2-22, Right.

Repairing Larger Breaks in Foil. When breaks are long, it is often better to use an insulated piece of wire to bridge the circuit gap. See Figure 2-23. The insulated wire eliminates the necessity of following the continual path of the foil in many instances. It also prevents contact with the bare conductor.

Once the repair is complete, a protective coating (varnish or approved equivalent) should again be applied to the PC board. The coating may be applied by brush or by spraying with an aerosol can. See Figure 2-24.

Repairing a Lifted Foil. A common problem caused from too much heat on a PC board is *foil separation*. See Figure 2-25. If the foil is merely lifted and not damaged, it should be reattached to the PC board with an appropriate glue. See Figure 2-26. Epoxy-type glues are probably best for adhering to both the copper foil and the insulated board. Apply the glue under the foil and firmly press the foil down onto the board. Hold the foil in place until the glue sets. If necessary, clean the foil and reapply any protective coating that may have been removed.

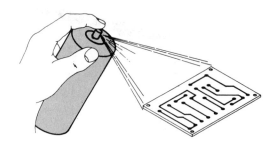

COATING APPLIED FROM AEROSOL CAN

COATING APPLIED BY BRUSH

Figure 2-24. A protective coating (varnish or approved equivalent) should be applied to the PC board once the repair is complete.

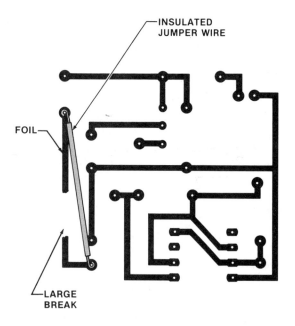

INSULATED JUMPER WIRE

FOIL

LARGE BREAK

NOT FOLLOWING TRACE (FOIL)

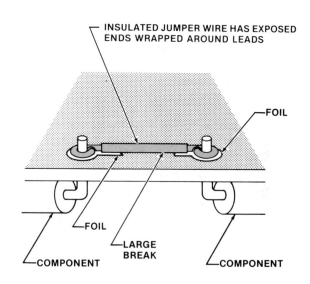

INSULATED JUMPER WIRE HAS EXPOSED ENDS WRAPPED AROUND LEADS

FOIL

FOIL

LARGE BREAK

COMPONENT

COMPONENT

FOLLOWING TRACE (FOIL)

Figure 2-23. For large breaks, soldering a piece of insulated jumper wire may be necessary to avoid shorts.

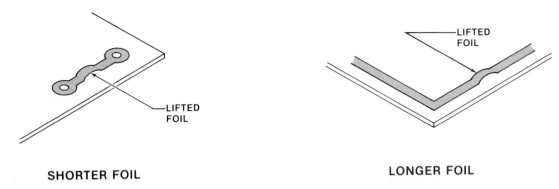

SHORTER FOIL

LONGER FOIL

Figure 2-25. Too much heat when soldering or desoldering components may cause the foil to lift away from the PC board.

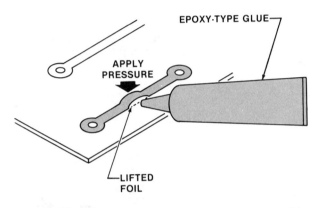

Figure 2-26. Epoxy-type glues and a reasonable amount of pressure should reattach the lifted foil without further repair.

SOLDERING PC BOARDS

The heat produced by a soldering iron can ruin a PC board. The critical factor is NOT the temperature of the soldering iron, but the wattage of the iron. To minimize the chance of overheating, it is best to use a *soldering pencil* with a rating of not more than 40 watts.

A functional soldering pencil station includes a variety of interchangeable soldering tips, a soldering pencil, a soldering pencil stand, and a sponge for cleaning the soldering pencil. Tips are selected according to the job to be performed. See Figure 2-27.

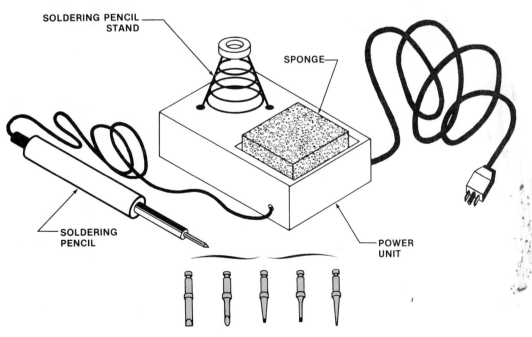

VARIOUS SOLDERING TIP CONFIGURATIONS FOR DIFFERENT JOBS

Figure 2-27. Soldering pencil stations are available with a variety of tips for specific jobs.

Preparing Soldering Pencil Tips

Most good soldering pencils have *coated tips* which require only that they are wiped clean after a sufficient warm-up time or heating time—usually 2–3 minutes. See Figure 2-28. However, some soldering tips are not coated and in time, oxidize. See Figure 2-29. Oxidation retards the transfer of heat and must be removed. Rubbing across the uncoated soldering tip with an abrasive cloth or light sandpaper removes the oxidation. See Figure 2-30. This procedure is important because an improperly heated solder joint, due to excessive oxidation on the soldering tip, results in a poor or defective connection.

NOTE: If the uncoated soldering tip is excessively oxidized, it may be necessary to use a file to remove the oxidation. The tip may also have to be removed and reshaped to its original shape or design.

To ensure that the oxidation does not return, a light coat of solder must be applied to the tip. See Figure 2-31. This process is called *tinning*, and it assures the maximum output of heat transfers from the soldering pencil.

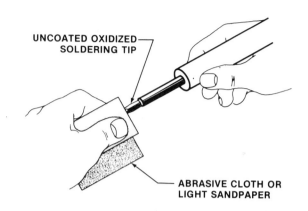

Figure 2-30. Rubbing an uncoated oxidized soldering tip with an abrasive cloth or light sandpaper cleans the tip.

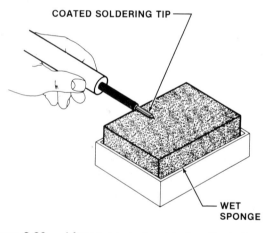

Figure 2-28. After a coated soldering tip is heated sufficiently, it can be wiped clean on a wet sponge.

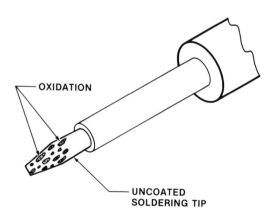

Figure 2-29. An uncoated soldering tip will oxidize.

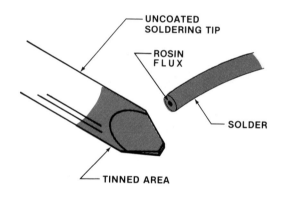

Figure 2-31. Once the oxidation is removed from an uncoated soldering tip, the tip must be tinned with a light coat of solder to prevent further oxidation.

Solder Selection

Use of a large diameter solder may result in damage to the circuit because of the excessive heat required for melting. A strong and lasting joint can be made with a small diameter, rosin-core solder that has a tin-to-lead ratio of approximately 60/40. See Figure 2-32.

CAUTION: Acid-core solder should never be used on electronic circuits because it results in corrosion leading to leakage and shorts.

The ideal solder (*eutectic*) is a combination of 63% tin and 37% lead. It melts at 360° F. Eutectic solder should be used with modular components.

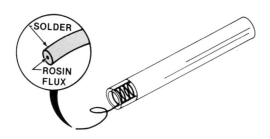

Figure 2-32. Rosin-core solder with a 60/40 tin-to-lead ratio should be used for soldering.

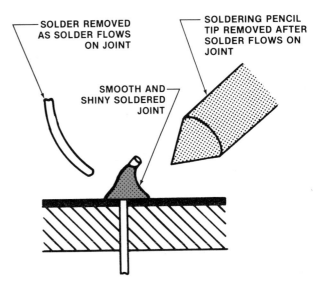

Figure 2-34. If both solder and soldering pencil tip are properly applied, solder will flow evenly across the joint.

Soldering Pencils

Good soldering skills are essential for repairing PC boards. Careless work creates unnecessary damage.

When soldering, the soldering pencil must heat the metal to solder-melting temperature BEFORE actual soldering can take place. The point of the soldering pencil tip should be held directly against one side of the component lead. See Figure 2-33. The solder should be applied from the opposite side. The solder-melting temperature is reached in a matter of 5 or 10 seconds. Therefore, the soldering pencil tip and the solder must be applied simultaneously.

CAUTION: Be sure to apply the solder to the opposite side of the component lead, not directly to the soldering pencil tip.

If the soldering procedure is done correctly, the solder will flow across the joint, firmly bonding the component lead to the PC board. See Figure 2-34.

Cold Solder Joints and Bridging

Two major problems created when soldering are the *cold solder joint* and *bridging*. The cold solder joint has a dull grainy texture. It exhibits a high resistance which adversely affects the operation of the circuit. A good solder joint is smooth and shiny. See Figure 2-35.

The problem of bridging is caused when solder joins, or bridges, two conducting paths that normally would be separated. See Figure 2-36. To avoid bridging, heat from the soldering pencil should be applied away from other conductors whenever possible. If bridging does occur, the solder may be drawn away by heat in some cases. In other cases, the solder must be removed, the surface scraped clean between the conductors, and the joint resoldered.

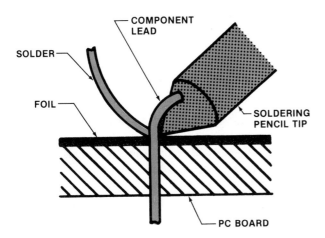

Figure 2-33. The point of the soldering pencil tip should be held to one side of the component lead while solder is applied to the opposite side.

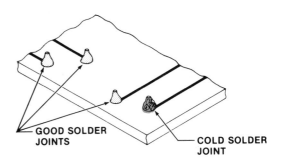

Figure 2-35. A cold solder joint has a dull grainy texture while good solder joints look smooth and shiny.

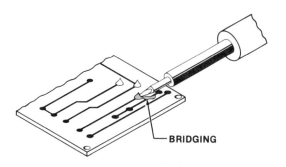

Figure 2-36. Two foils are bridged by solder. Heat from the soldering pencil was too close to the second foil or trace, and the solder was drawn to the second foil.

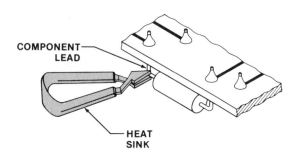

Figure 2-37. A typical heat sink applied to a component lead protects the component from heat damage.

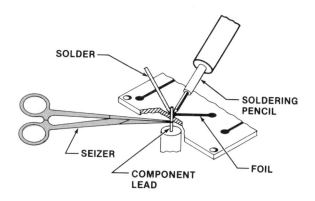

Figure 2-38. A seizer is used where space is critical and a small heat sink is necessary.

Heat Sinks

Heat sinks are mechanical devices used to draw heat away from components and protect them from heat damage. There are a variety of heat sinks available. A typical heat sink is shown in Figure 2-37. When space is critical or when a small component is involved, a special heat sink called a *seizer* is used. See Figure 2-38.

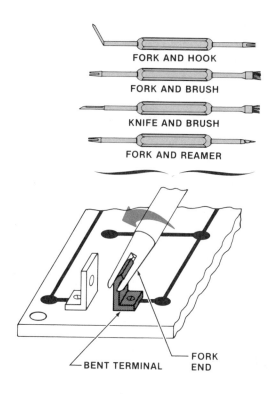

Figure 2-39. Soldering aids are instruments designed to loop wires, hook wires, deburr holes, and brush away hot solder.

Soldering Aids

Soldering aids are instruments designed specifically for PC board repair. See Figure 2-39. They are made of non-magnetic steel so solder will not adhere to them. The fork end straddles wire for looping and guiding. The hook end is used to hook wires. The reamer end cleans lug holes and is also used for deburring. The knife end removes surplus solder, and the brush end cleans solder connections.

Specialized Desoldering Tips

Straight-line, multi-terminal connections can be removed by heating each solder connection individually and by brushing away the melted solder. However, a less time-consuming method to expedite desoldering is to apply a *bar-type desoldering tip* across all the connections at once, and remove the component in a single action. See Figure 2-40.

For dual in-line sockets and directly-mounted, integrated circuits, the *dual in-line desoldering tip* is suitable. See Figure 2-41. For circular configurations, the *cup-type desoldering tip* is appropriate. See Figure 2-42.

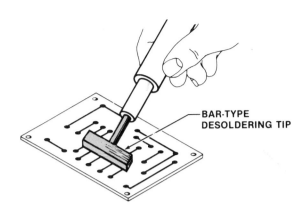

Figure 2-40. A bar-type desoldering tip is used to remove straight-line, multi-terminal connections.

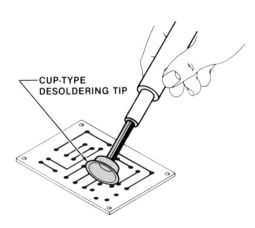

Figure 2-42. A cup-type desoldering tip is used to remove transistor and integrated circuits in a circular configuration.

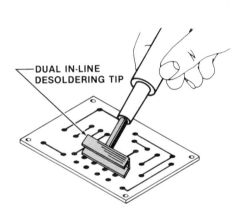

Figure 2-41. A dual in-line desoldering tip is used to remove directly-mounted, integrated circuits.

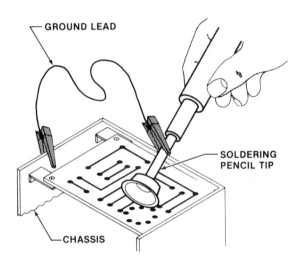

Figure 2-43. In emergency situations where a PC board cannot be removed from the circuitry, it is good practice to place a ground lead from the soldering pencil tip to the chassis to avoid current leakages from the soldering pencil.

Chassis Mounted PC Board— Component Removal

WARNING: The following procedure is used primarily in an emergency situation and should be avoided whenever possible. It is more convenient and always safer to completely remove the PC board and service it on an insulated surface, or in a PC board holding jig.

Unless the soldering pencil is transformer-isolated, it is good practice to connect a ground lead from the tip of the soldering pencil to the chassis or to the PC board ground. This prevents damage to transistors and other parts due to current leakage from the soldering pencil. See Figure 2-43.

It may be necessary to replace a defective resistor or capacitor without removing the PC board from the chassis. If the leads extending from the defective component are long enough for a replacement part to be soldered to them, they may be cut where they enter the defective component. If the leads are not long, cut the component in half. See "A" of Figure 2-44. Remove the excess material by crushing it to take advantage of the lead length within the component. See "B" of Figure 2-44. When enough lead length is available, loop the old leads and solder the replacement component to the old leads. See "C" and "D" of Figure 2-44.

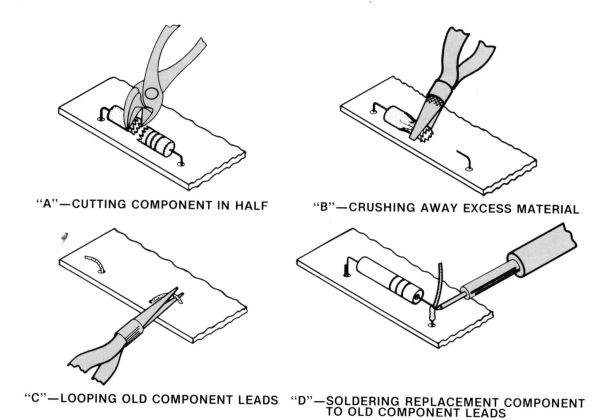

"A"—CUTTING COMPONENT IN HALF   "B"—CRUSHING AWAY EXCESS MATERIAL

"C"—LOOPING OLD COMPONENT LEADS   "D"—SOLDERING REPLACEMENT COMPONENT
TO OLD COMPONENT LEADS

Figure 2-44. In emergency situations, components may be cut from a board and replaced quickly using a crush-and-cut procedure.

Chapter 2 - Review Questions

1. What is a printed circuit board?
2. Name the major components of a PC board.
3. What is an edge card connector?
4. How are components located on a PC board?
5. What important points must be observed when taking electrical measurements on a PC board?
6. When removing a PC board, what safety precautions should be observed?
7. How should a PC board be held while being serviced?
8. What methods are used to remove solder from a PC board?
9. How can an open circuit be isolated on a PC board?
10. Describe the procedure for repairing large breaks in a PC foil.
11. Describe the procedure for repairing a lifted foil.
12. Describe the procedure for repairing small breaks in a foil.
13. What is the most important factor in selecting the proper soldering pencil?
14. What is the best type of solder for PC board repair?
15. Name and describe the two major problems created when soldering.
16. What are heat sinks?
17. What type of soldering aids are available?
18. What types of specialized desoldering tips are available?
19. How are emergency PC board repairs made?
20. How can a PC board be protected from current leakage?

3 SEMICONDUCTOR DIODES

Diodes are components that have the unique ability of allowing current to pass through them in only one direction. When the diode was first developed, it was made of a metal plate coated with a selenium material. Semiconductor diodes, usually referred to simply as diodes, are the most widely used diodes now because of their small size and weight. See Figure 3-1. The semiconductor diode has unique properties because of the material from which it is made. The two most common materials used are silicon and germanium. Neither element conducts electricity well in its pure form, but by adding certain impurities (See DOPING), they become very conductive when a voltage of the proper polarity is applied.

Application of the proper polarity to a semiconductor diode is called a forward biased voltage, or *forward bias*. Forward bias results in *forward current*. When the opposite polarity is applied, it is called a reversed biased voltage, or *reverse bias*. Reverse bias results in a *reverse current* which should be close to zero or very small—usually a millionth of an ampere (μA).

A diode has a relatively low resistance in the forward bias direction, and a considerably greater resistance in the reverse bias direction. A schematic symbol representing a semiconductor diode is composed of two parts: a straight line which represents the *cathode*, and a triangle which represents the *anode*. See Figure 3-2. Electrons flow from cathode to anode, or against the triangle. The diode in Figure 3-2 is shown in forward bias. If a negative polarity is applied to the anode and a positive polarity is applied to the cathode, the diode would be in reverse bias and no current would flow. See Figure 3-3.

NOTE: In this text, current flow is based on the electron theory which states that current flows from negative to positive. Conventional current flow is based on current flow from positive to negative. The use of electron theory and electron current flow will be used throughout the text.

Key Words

Anode	Forward Bias	Pulsating DC
Avalanche Current	Forward Breakover Voltage	Rectifier
Avalanche Diode	Forward Current	Reverse Bias
Biased Clamping Circuit	Forward Operating Current	Reverse Current
Carriers	Half-Wave Rectifier	Series Regulator
Cathode	Heat Sink	Shunt Regulator
Clamper	Holes	Silicon Grease
DC Restorer	Intermittents	Temperature Coefficient
Depletion Region	N-Type Material	Thermal Contact
Derated	Operating Characteristic Curve	Voltage Limiter
Diodes	P-Type Material	Voltage Regulator
Diode Clamping Circuit	Peak Inverse Voltage (PIV)	Zener Diode
Doping		

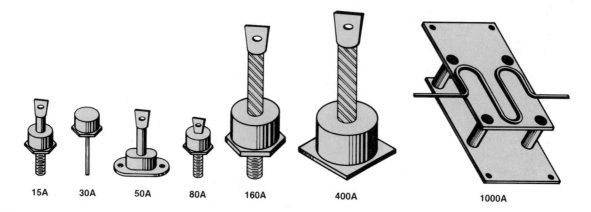

| 15A | 30A | 50A | 80A | 160A | 400A | 1000A |

Figure 3-1. Semiconductor diodes are relatively small and lightweight. They have the unique property of allowing current to pass through them in only one direction. The size of a diode increases with an increase in amperage rating.

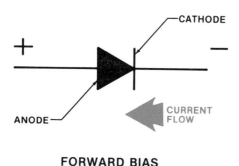

FORWARD BIAS

Figure 3-2. The schematic symbol for a semiconductor diode is a vertical line for the cathode and a triangle for the anode. Electrons flow from the cathode to the anode, or against the triangle when in forward bias.

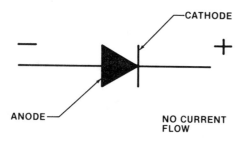

REVERSE BIAS

Figure 3-3. No current will flow when the polarity on the diode is reversed (reverse bias).

RECTIFIERS

Many electronic devices require DC (direct current) power for operation. However, AC (alternating current) power is generally the most readily available

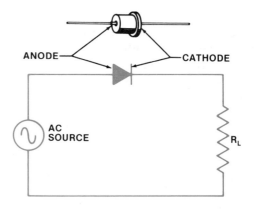

RECTIFIER CIRCUIT

Figure 3-4. The half-wave rectifier is the simplest form of a rectifier circuit.

power. Therefore, AC power must be converted to DC power before it can be used in many applications. Circuits that provide this conversion are called *rectifiers*, and these circuits use semiconductor diodes.

The simplest form of a rectifier circuit is the *half-wave rectifier*. See Figure 3-4. It is composed of a load resistor R_L combined with a diode across an AC source. This circuit can affect the AC voltage by cutting the output voltage in half, or by rectifying it. See Figure 3-5. Since the signal now travels in only one direction, it is classified as *pulsating DC*. A half-wave rectifier is often mounted on a PC board. See Figure 3-6. The half-wave rectifier is only one type of power supply circuit in which diodes are used.

DIODE MARKINGS

There are several ways in which manufacturers mark diodes to indicate the cathode and the anode. See

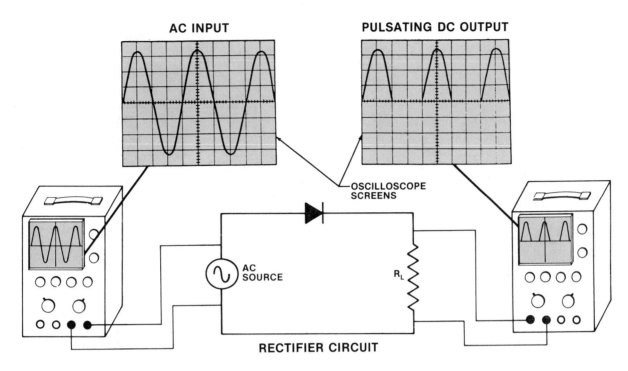

Figure 3-5. The half-wave rectifier is a combination of a load resistor and a diode across an AC source. When operating properly, the output voltage will be cut in half, or rectified. Since the signal now travels in only one direction, it is classified as pulsating DC.

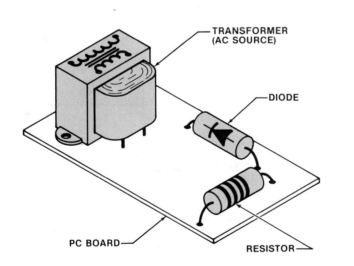

Figure 3-6. A half-wave rectifier may be mounted on a PC board with a transformer as the AC source.

Figure 3-7. Diodes may be plainly marked with the schematic symbol, or there may be a band at one end to indicate the cathode. Some manufacturers use the shape of the diode package to indicate the cathode end. Usually the cathode end is beveled or enlarged to ensure proper connection. When unsure, it is best to check the polarity of a diode with an ohmmeter.

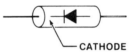

DIODE MARKED WITH SCHEMATIC SYMBOL

CATHODE MARKED WITH A BAND

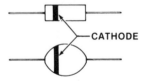

CATHODE END PHYSICALLY LARGER

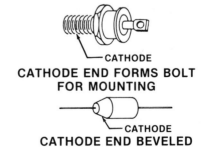

CATHODE END FORMS BOLT FOR MOUNTING

CATHODE END BEVELED

Figure 3-7. Manufacturers use a variety of methods to indicate the cathode end of a diode.

CONDUCTOR	SEMICONDUCTOR	INSULATOR
LOW RESISTANCE		HIGH RESISTANCE

Figure 3-8. A resistance scale shows the relationship between conductors, semiconductors, and insulators.

UNDERSTANDING SEMICONDUCTOR MATERIALS

Resistance to current flow from semiconductor materials falls somewhere between the low resistance offered by a conductor, and the high resistance offered by an insulator. A resistance scale shows the resistance of semiconductor materials in relation to conductors and insulators. See Figure 3-8.

Doping

The basic material used in most semiconductor devices is either germanium or silicon. In their natural state, germanium and silicon are pure crystals. These pure crystals do not have enough free electrons to support a significant current flow. To prepare these crystals for use as a semiconductor device, their structure must be altered to permit significant current flow.

Doping is the process by which the crystal structure is altered. In doping, some of the atoms in the crystal are replaced with atoms from other elements. The addition of new atoms in the crystal creates *N-type material* and *P-type material*.

N-Type Material. N-type material is created by doping a region of a crystal with atoms from an element that has more electrons in its outer shell than the crystal. Adding these atoms to the crystal results in the possibility of more free electrons. See Figure 3-9. Since electrons have a negative charge, the doped region is called N-type material.

As in any conductor, free electrons support current flow. When voltage is applied to N-type material, current flows from negative to positive through the crystal. See Figure 3-10. The free electrons help move current and are called *carriers*.

Some elements commonly used for creating N-type material are arsenic, bismuth, and antimony. The quantity of doping material used generally ranges from a few parts per billion to a few parts per million. By controlling even these small quantities of impurities in a crystal, the manufacturer can control the operating characteristics of the semiconductor.

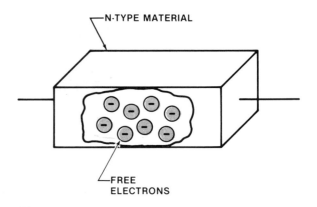

Figure 3-9. N-Type material is created by doping the basic crystal with a material that provides extra, or free electrons.

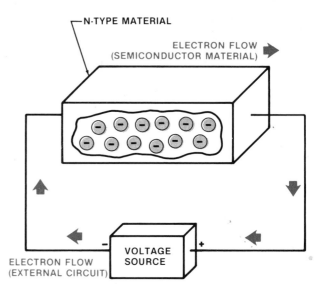

Figure 3-10. When a voltage is applied to N-Type material, current flows from negative to positive through the help of free electrons called carriers.

P-Type Material. To create P-type material, a crystal is doped with atoms from an element that has fewer electrons in its outer shell than the natural crystal. This combination creates empty spaces in the crystalline structure. The missing electrons in the crystal structure are called *holes*. The holes are represented as positive charges. See Figure 3-11.

Typical elements used for doping a crystal to create P-type material are gallium, boron, and indium. In P-type material, the holes act as carriers. When voltage is applied, the holes are filled with free electrons as the free electrons move from negative potential to positive potential through the crystal. See Figure 3-12.

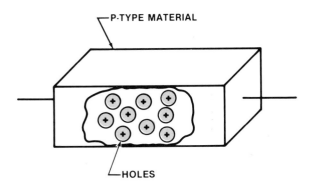

Figure 3-11. P-Type material is created by doping the basic crystal with a material that has a deficiency of electrons. These missing electrons are called holes and are represented by a positive charge.

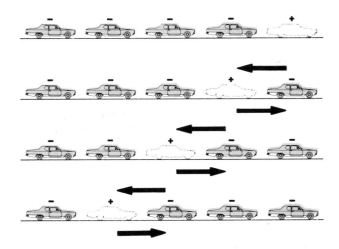

Figure 3-13. In the sequence of events in this parking lot analogy, the hole appears to move across the parking lot.

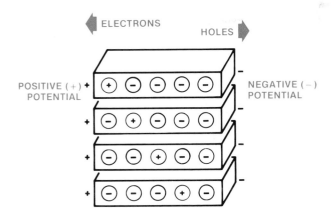

Figure 3-12. Movement of the electrons from one hole to the next makes the holes appear to move in opposite direction of the electrons.

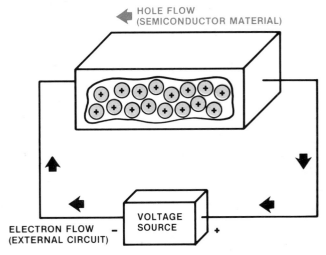

Figure 3-14. The direction of hole flow inside P-Type material appears to move from positive to negative, while the current outside the P-Type material flows from negative to positive.

Movement of the electrons from one hole to the next makes the holes appear to move in the opposite direction. Hole flow is equal to and opposite of electron flow.

The concept of hole flow may be difficult to grasp at first. To help visualize this concept, the parking lot analogy can be used. See Figure 3-13. In this illustration, a car has left the parking lot and has created a hole (empty space). If the car directly behind this hole moves forward, then one space is filled and another hole is created. As each car proceeds to the next position, the hole appears to move backwards while the cars move forward. This same concept applies to the P-type material used in the semiconductor. The direction of hole flow inside P-type material appears to move from positive to negative. The direction of electrons outside the P-type material flow from negative to positive. See Figure 3-14.

Depletion Region

Diodes have the ability to block current flow in one direction and pass current in the opposite direction. This is made possible by the doping process, which creates N-type material (region) and P-type material. At the junction of the two materials, the P-type and N-type materials exchange carriers, creating a thin zone called the *depletion region*. See Figure 3-15. Because the depletion region is very thin, it responds rapidly to voltage changes.

When voltage is applied to the diode, the action occurring in the depletion region either blocks current flow, or passes current.

Forward Bias

Forward bias exists when voltage is applied to a solid-state device in such a way as to allow the device to conduct easily. When a diode is in forward bias, the polarity of the power source causes the free electrons and holes to move toward the depletion region. See Figure 3-16. The carriers bridge the depletion region and cause it to close. In this condition, carriers are available from one end of the diode to the other, allowing normal current flow with little resistance.

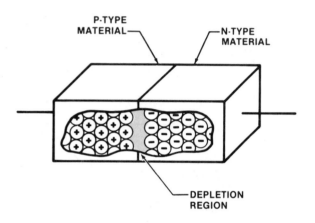

Figure 3-15. At the junction of the P-Type and N-Type material, the materials exchange carriers, and a thin neutral area called a depletion region is formed.

Reverse Bias

Reverse bias exists when voltage is applied to a device in such a way that it causes the device to act as an insulator. When reverse bias is applied to a diode, the polarity of the power source causes the free electrons and holes to move away from the depletion region. See Figure 3-17. This action increases the size of the depletion region and thus, blocks current flow.

OPERATING CHARACTERISTIC CURVES

An *operating characteristic curve* shows the relationship between voltage and current for a typical diode. Manufacturers often supply operating characteristic curves with the diodes. These curves show how diodes operate. They also give information concerning the specifications of the diodes.

Diode operating characteristic curves are graphs. The axes of the graph provides specific information. See Figure 3-18. The horizontal axis of the characteristic graph represents the voltage applied to a diode. The vertical axis represents the current flow through the diode. The intersection of the horizontal and vertical axes is called the origin. The origin represents zero on both axes.

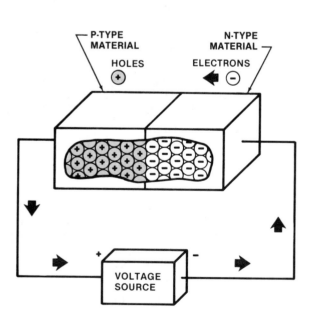

Figure 3-16. When a diode is in forward bias, the polarity of the power source moves the free electrons and holes toward the depletion region, causing it to close. Current flows with little resistance.

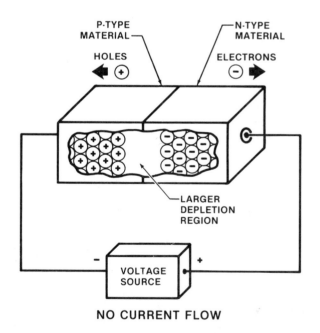

Figure 3-17. When a diode is in reverse bias, the polarity of the power source moves the free electrons and holes away from the depletion region, making the region larger. Current does not flow because of the high resistance.

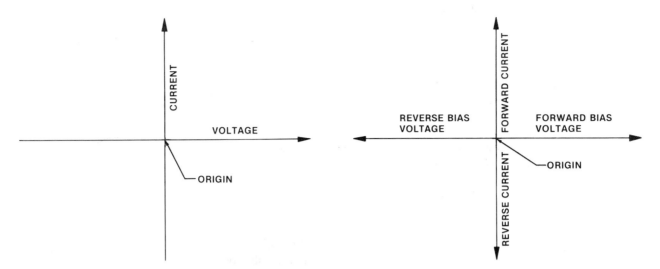

Figure 3-18. The intersection of the horizontal and vertical axes is called the origin. The vertical axis represents current and the horizontal axis represents voltage.

Figure 3-19. The horizontal axis to the right of the origin represents forward bias voltage, while the horizontal axis to the left of the origin represents reverse bias voltage. The vertical line above the origin represents forward current due to forward bias. The vertical line below the origin represents reverse current due to reverse bias.

The horizontal line to the right of the origin represents forward bias voltage. The horizontal line to the left of the origin represents reverse bias voltage. The vertical line above the origin represents forward current due to forward bias. The vertical line below the origin represents reverse current due to reverse bias. See Figure 3-19.

Forward Operating Current

When voltage is applied to a diode in the forward direction, current will not flow until the depletion region is closed. This voltage is often called the *forward breakover voltage*. Only a few tenths of a volt is needed for the carriers to bridge the depletion region (about 0.6 volts for silicon and 0.3 volts for germanium). On the operating characteristic curve, the depletion region is indicated by the short flat line that runs parallel to the horizontal axis. See Figure 3-20.

Once the depletion region is closed, resistance across the diode is very low and current flow increases rapidly. If the forward bias voltage remains, the diode continues to pass current freely. On the operating characteristic curve, the region of rapid current rise is called the *forward operating current*. See Figure 3-20. When the forward operating current is high, a heat sink may be necessary to dissipate the heat generated.

Depending upon the physical size and rating of the diode, typical forward operating current ranges from a few milliamps (mA) to several hundred amps (A). The value of the operating current depends on the type

of diode. This operating current rating indicates the maximum continuous current level a diode can safely handle.

CAUTION: A replacement diode should have at least the same forward operating current rating as the original diode.

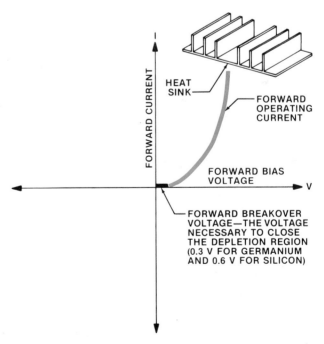

Figure 3-20. On an operating characteristic curve, the region of rapid current increase is called the forward operating current which, when it is quite high, may require the use of a heat sink.

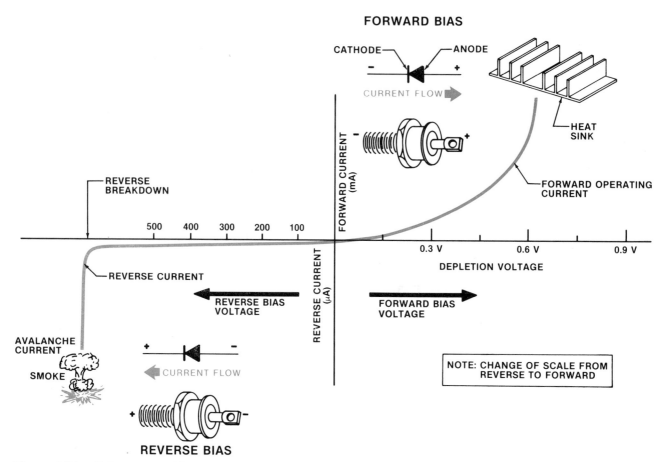

Figure 3-21. If the reverse bias applied to a diode exceeds its peak inverse rating, the diode breaks down and passes current freely, causing the diode to be destroyed.

Peak Inverse Voltage (PIV)

Diodes are also rated for the maximum reverse bias voltage they can withstand. This rating is called *peak inverse voltage (PIV)*. The PIV ratings for most diodes used in industry range from a few volts to several thousand volts. If the reverse bias applied to the diode exceeds its PIV rating, the diode will break down and pass current freely. Current passed in this manner is called *avalanche current*, which can destroy diodes. See Figure 3-21. To avoid avalanche, diodes with the correct rating must be used.

CAUTION: A replacement diode must have an equal or greater PIV rating than the original diode.

TESTING DIODES

In most cases, diodes are tested with an ohmmeter. The polarity of the diode can be determined with an ohmmeter, and the diode can also be checked for opens and shorts.

CAUTION: Because ohmmeters vary somewhat from manufacturer to manufacturer, consult the operator's manual before proceeding to make any tests.

Diode Polarity

An ohmmeter can be used to determine which end of the diode is the cathode and which end is the anode. This is possible because the ohmmeter is a voltage source with a definite polarity.

Externally, the polarity of the ohmmeter may be marked positive (+) and negative (−). It may also be identified by a color coding system—usually red for positive and black for negative. See Figure 3-22, left. Internally, however, the voltage source, or battery, actually determines the external polarity. See Figure 3-22, right. The determination of polarity is important.

The forward and reverse bias of an unknown diode can be determined if the diode is placed between known polarities one way, and then placed in the opposite direction. The diode will indicate a low resistance in forward bias and a high resistance in reverse bias. Since the polarity of the source is known, the end connected to the negative lead during forward bias must be the cathode, and the end connected to the positive lead must be the anode. See Figure 3-23.

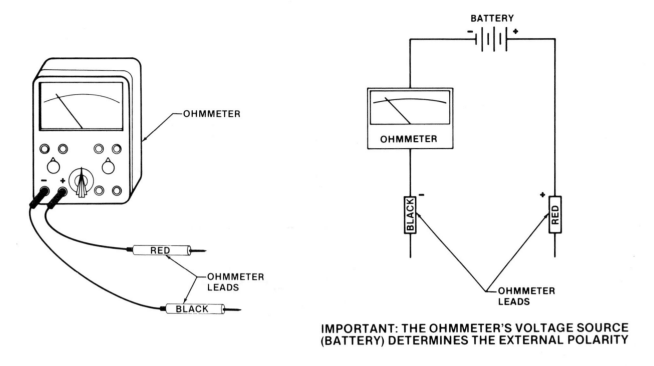

IMPORTANT: THE OHMMETER'S VOLTAGE SOURCE
(BATTERY) DETERMINES THE EXTERNAL POLARITY

Figure 3-22. A standard ohmmeter is shown on the left while its equivalent electrical circuit is shown on the right.

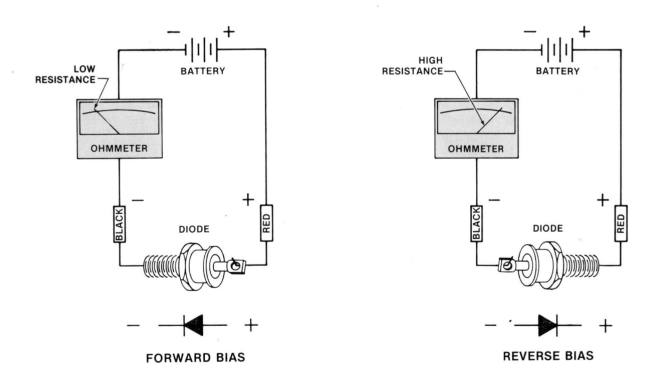

Figure 3-23. A diode indicates a low resistance in forward bias and a high resistance in reverse bias. Since the polarity of the source is known, the end connected to the negative lead during forward bias is the cathode end.

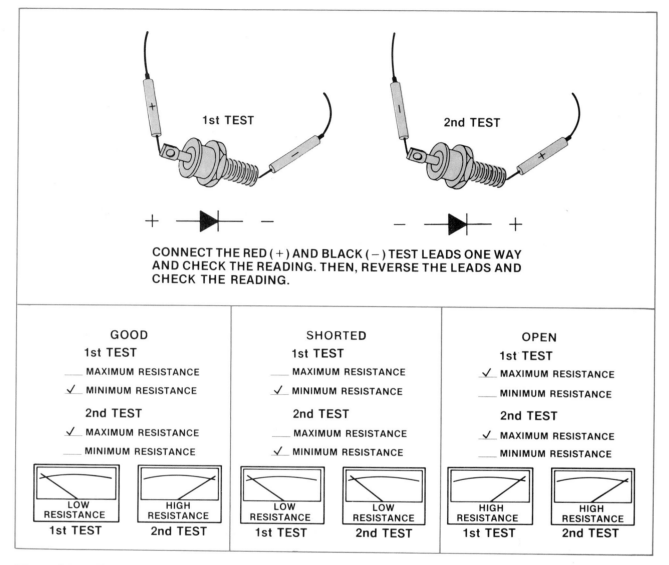

Figure 3-24. Two tests should be conducted on a diode to determine its operating condition. These tests can determine if a diode is shorted, open, or good.

Testing Diodes for Opens and Shorts

Diodes are very reliable; however, they are not indestructible. High voltages, improper connections, and overheating can damage a diode. The electrician or technician may be responsible for determining the condition of the diode. Opens and shorts are the two most common diode problems. The ohmmeter is very helpful in determining the condition of the diode.

A definite procedure is used when testing a diode. See Figure 3-24. A diode is usually tested twice. If the diode is good, there is a low resistance reading in one direction and a high resistance reading in the opposite direction. If the diode is shorted, a low resistance reading registers in both directions. If the diode is

open, there is a high resistance reading in both directions. If a diode is open or shorted, it must be replaced.

Ohmmeter Precautions

Always read the operator's manual on the multimeter to determine the exact polarity of the source voltage in the ohmmeter function. A red lead is not always positive, and a black lead is not always negative. It is up to the discretion of the multimeter manufacturer to determine internal polarities. Therefore, the operator must always check the polarity before using an ohmmeter as an established polarity.

CAUTION: When using an ohmmeter, be sure that no power is applied to the diode in the circuit. Also, make sure that at least one lead of the diode is disconnected from the circuit to avoid interference from any parallel circuitry.

DIODE INSTALLATION AND SERVICE

To ensure that diodes will operate properly, these points must be observed when installing or servicing a diode:

1. Voltages: Observe voltage specifications. Check the power line voltage to make sure the diode is properly rated.

2. Current: Do not overload diodes, even momentarily. Double check circuits, polarities, component sizes, and wiring BEFORE installation.

3. Mounting: Stud-mounted diodes must be fastened properly and securely to their heat sinks to assure good heat flow. See Figure 3-25. Do not overtighten the stud or it will stretch or strip. The best way to install diodes in heat sinks is to use a torque wrench. It is also advisable to follow manufacturer's specifications.

4. Heat sinks: Carefully observe the manufacturer's recommended use of heat sinks. See Figure 3-26. If heat cannot escape, damage is likely. Be sure air can circulate around the device. Also, watch for excessively high temperatures caused by other components such as nearby lamps, motors, or heaters.

5. Silicon grease: When diodes are mounted on heat sinks, silicon grease is often applied to the surfaces of the metal where contact is made. Silicon grease increases the thermal conductivity of the metal and allows heat to transfer more rapidly.

6. Soldering: The amount of heat developed by soldering irons and pencils can damage semiconductor devices. Care must be exercised when semiconductor devices, such as diodes, are removed or replaced when using soldering equipment. In most cases, heat-sensitive components can be protected through the use of a heat sink. See Figure 3-27. This small metallic clamp is designed to draw heat away from the body of the semiconductor device. When a heat sink is not available, an alligator clip may be substituted. See Figure 3-28. A pair of long-nose pliers may also be substituted as a heat sink. It may, however, require a rubber band or wire wrapped around the handles to maintain good thermal contact. See Figure 3-29.

CAUTION: Heat sinks should not be removed immediately. In most cases, allow 30–45 seconds before removing the heat sink.

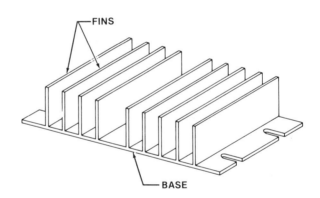

Figure 3-26. To be effective, heat sinks must be properly installed. The manufacturer's recommended uses should be considered.

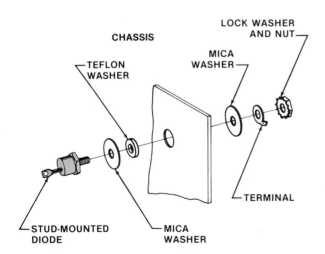

Figure 3-25. It is often desirable to mount the cathode side of a large stud-mounted diode on the chassis. The chassis then can be used as an effective heat sink to conduct excess heat away from the diode.

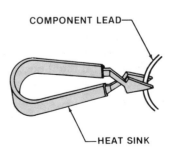

Figure 3-27. The heat sink is a small metallic clamp designed to draw heat away from the body of a semiconductor device.

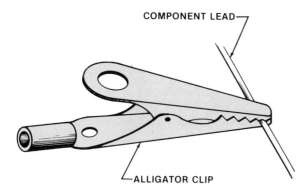

Figure 3-28. An alligator clip is a good substitute for a heat sink.

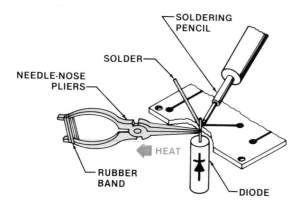

Figure 3-29. When needle-nose pliers are used as a heat sink, a rubber band or wire may have to be wrapped around the handles to maintain good thermal contact.

DIODE POWER CAPACITY AND DERATING

The maximum current at which a diode may operate in a normal environment is limited primarily by the temperature rise at the PN junction. It is also limited by the size of the heat sink provided to dissipate the heat. The power capacity of most diodes is rated at an operating temperature of 25 °C, or approximately room temperature. As the ambient temperature increases beyond this point, the diode can no longer dissipate as much heat and must be *derated*. Derating a diode means to operate it at less than maximum operating current. Derating tables are available to determine this value.

When it is necessary to operate a diode at or close to the maximum current rating, heat sinks must be used. Heat sink tables are available to determine the proper size for a particular job and should be consulted.

PRACTICAL APPLICATIONS OF DIODES

Uses for a diode vary. By connecting the diode in series, it can stop current from entering another component. By connecting the diode in parallel, it can bypass potentially damaging currents. By using series-parallel arrangements, diodes can direct current as required.

Diode Current Protection for Meter Movement

Low-voltage silicon diodes can be used to protect DC meter movements from excessive currents. The circuit has two diodes that are used as protective shunts. See Figure 3-30. Shunts allow excess current to bypass the meter movement by offering a path of less resistance.

When the voltage across the meter movement exceeds 0.6 volts in either direction, a diode is forward biased and shunts most of the current around the meter, thereby protecting it.

For a typical multimeter with a meter movement resistance of 1,200 ohms and a full-scale current rating of 50 μA, the rectifiers introduce less than 1% error into the meter reading. At the same time, the meter movement current is limited to less than 1 mA for a one-ampere fault current. This value is well below the destructive value of most meter movements.

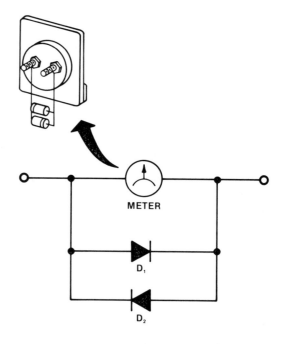

Figure 3-30. Low-voltage silicon diodes can be used to protect DC meter movements where heavy fault currents are possible.

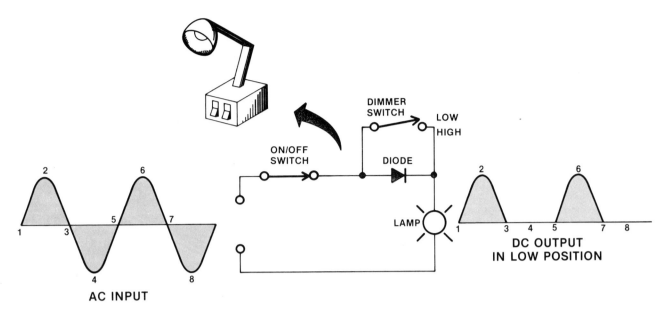

AC INPUT

NOTE: THE NUMBERS ON THE WAVEFORMS ARE POINTS WHICH ARE USED ON-LY TO HELP ILLUSTRATE THE REDUCTION IN CURRENT OUTPUT. THEY ARE NOT CURRENT VALUES.

Figure 3-31. A diode may be used as a half-wave rectifier to reduce the amount of current flowing through the load by one-half.

Diode as a Light Dimmer

A diode may be used as a half-wave rectifier to reduce the amount of current flowing through the load by one-half. See Figure 3-31. This circuit is typical of two-position light dimmers and high intensity lamps that can be switched to bright or dim. When the dimmer switch is closed, or in the high position, the diode is shorted out, allowing full current and full brightness. When the dimmer switch is open, or in the low position, the diode is placed in series with the lamp, blocking one-half the current. The intensity of the lamp is therefore reduced.

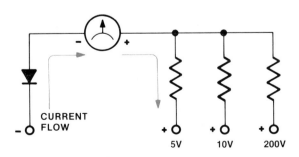

Figure 3-32. To protect a meter movement from reverse polarity, a diode may be placed in series with the meter movement so that current flows through it in only one direction.

Diode Polarity Protection for Meter Movement

If the leads of a voltmeter are accidently placed across a voltage source in reverse polarity (reverse direction), the meter movement may be damaged. Damage to the meter movement occurs when the meter pointer pings against the stops with such force that the meter pointer is bent or knocked out of calibration. To avoid reverse polarity on a meter movement, a diode may be placed in series with the meter movement. Current then flows through it in only one direction. See Figure 3-32.

Diode Clipping or Limiting Circuits

In certain electronic circuits, it is necessary to limit the amount of voltage present in the circuit. This is particularly true where large voltage changes are expected. A diode clipping, or limiting circuit (*voltage limiter*), is used for this type of protection.

A simple voltage limiter can be constructed from a resistor and a reverse biased diode. See Figure 3-33. The bias voltage and the voltage required to forward bias the diode determine the maximum positive peak output of the voltage limiter.

In the circuit shown in Figure 3-33, a bias voltage of 3 volts is connected to the positive terminal of the bias voltage source. The bias voltage source in turn is connected to the cathode of the diode. This polarity on the diode causes the diode to be reverse biased and

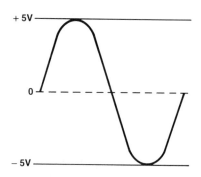
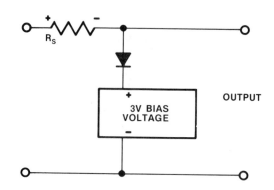

Figure 3-33. A voltage limiter is used to limit the amount of voltage in a circuit. A simple voltage limiter can be constructed with a resistor and a reverse-biased diode.

no current flows in the circuit. To obtain current flow in this circuit, a positive signal of 3.6 volts or greater must be applied to the circuit. A voltage of 3.0 is necessary to neutralize the reverse bias, and a voltage of 0.6 is necessary to make the diode conduct. For any voltage less than 3.6, the diode acts as an open switch with the entire output across the open circuit.

However, when a signal of 3.6 volts or greater is applied, the diode conducts and effectively shorts out the output. The positive output of that half cycle is then limited to the bias voltage of 3.6 volts. See Figure 3-34. The entire negative half of the cycle appears across the output since the diode is an open circuit.

Diode Clamping Circuits

A *diode clamping circuit* is a circuit that holds the voltage or current to a specific level. The component commonly used as a *clamper* (clamping device) is the diode. During the clamping process, the diode should not significantly change the shape of a waveform.

The main function of a clamper is to provide a DC reference level for a signal voltage. The clamper effectively raises or lowers the positive or negative peaks of a signal to zero or some other desired reference level. See Figure 3-35. Clampers are also called *DC restorers*.

A diode clamping circuit can be shown in a schematic. See Figure 3-36. Segments A through E represent the first AC cycle appearing at the input terminals. Segments A1 through E1 are the corresponding output voltages. The solid-line portion of each segment shows only the part of the cycle that is affected at that point in time.

The portion of the cycle shown in segment A charges capacitor C through forward-biased diode D. The charge current flows in the direction of the arrow.

With the polarity shown, capacitor C charges to the maximum positive value of the AC cycle. As the capacitor charges, the capacitor voltage is always equal to and opposite of the AC input signal voltage. The opposite voltages cancel each other, producing a zero volt output. See Segment A1 of Figure 3-36.

As the positive portion of the cycle decreases, the diode becomes reverse bias. See Figure 3-36, Segment B. The capacitor then tries to discharge some of its stored potential through resistor R. However, the RC time constant is always long compared to the time duration of the input voltage. With resistor R large enough, the capacitor loses only a small amount of its charge over a complete AC cycle.

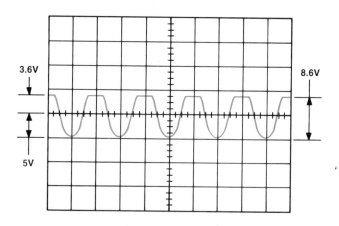

Figure 3-34. The bias voltage plus the voltage required to forward bias the diode determine the maximum positive peak output of a voltage limiter.

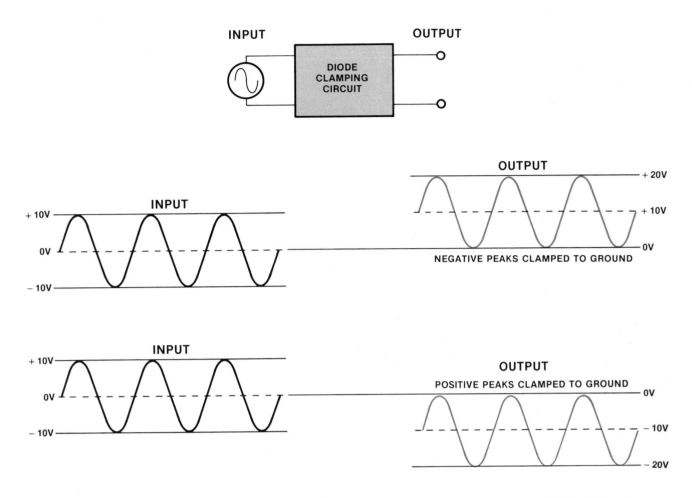

Figure 3-35. A diode clamping circuit is a circuit that holds the voltage or current to a specific level. The clamper effectively raises or lowers the positive or negative peaks of a signal to zero or some other desired reference level.

As the AC input signal voltage continues to decrease (Segment B), the capacitor acts like a battery and opposes the signal voltage. The negative voltage shown at the output (Segment B1) represents the difference between the charge on the capacitor and the decreasing AC voltage. Finally, the point is reached in Segment B where the signal returns to zero. The corresponding voltage segment B1 only represents the charge on the capacitor, and ranges from 0 volts to −10 volts.

In Segments C and D, the AC input signal voltage is shown in its negative polarity region. All of the circuit conditions described for Segment B still apply. However, instead of opposing, the capacitor voltage now aids the negative-going signal voltage. When the full negative value of the signal is reached, the corresponding output is shown in segments C1 and D1.

NOTE: The output voltage is twice the value of either the positive or negative alternation at the input.

In Segment D, the signal returns once again to zero. The corresponding output voltage is the charge on the capacitor. It ranges from −20 volts to −10 volts.

The cycle repeats as shown in Segment A. However, the output waveform shown in Segment A1 occurs only on the first input cycle. The corresponding output waveform for all later cycles is the one in Segment B1.

NOTE: The circuit of Figure 3-36 clamps the positive peaks of the input signal to ground. A circuit that clamps the negative peaks to ground is made by reversing the diode shown in Figure 3-36.

A diode clamping circuit can be used when a reference level other than zero is needed. A clamper can provide an off-ground reference level. See Figure 3-37. This circuit uses a battery to reverse bias the diode and is called a *biased clamping circuit*. It is similar to the diode clamping circuit. The difference is that the biased clamping circuit produces an output reference level equal to the battery voltage.

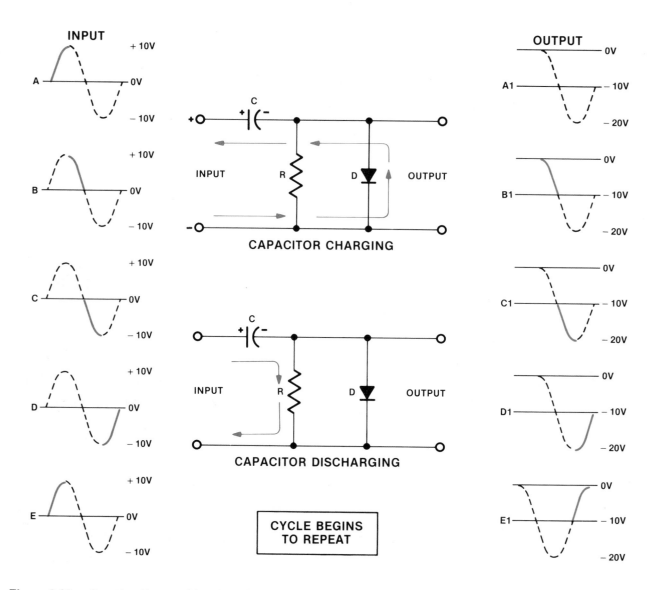

Figure 3-36. By using the combination of a diode-capacitor network, the positive peaks of the input signal can be clamped to ground.

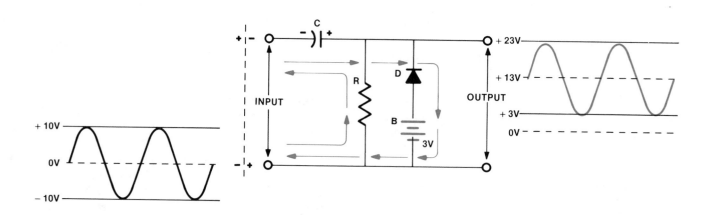

Figure 3-37. By using a bias voltage, a diode clamping circuit can be used for a reference other than zero. This circuit is called a biased clamping circuit.

The battery shown in Figure 3-37 applies 3 volts of forward bias to the diode. When the diode conducts, capacitor C charges to about 3 volts. The first negative alternation at the input also causes the capacitor to charge. However, on the negative alternations, the capacitor voltage opposes the input voltage. Therefore, the output voltage cannot be less than +3 volts.

On the positive alternations, the capacitor voltage aids the input voltage. Therefore, the maximum positive voltage is equal to the peak-to-peak value of the input voltage, plus the bias voltage. In this case, the maximum positive voltage is 23 volts.

ZENER DIODE

The *zener diode* looks similar to other silicon diodes. See Figure 3-38. Its purpose is to act as a *voltage regulator* either by itself or in conjunction with other semiconductor devices. In a schematic, the zener diode symbol differs from a standard diode symbol in that the normally vertical cathode line is bent slightly at each end.

The zener diode is unique because it is most often used to conduct current under reverse bias conditions. Standard diodes usually conduct in the forward bias condition and can be destroyed if the reverse voltage or bias is exceeded. Because the zener diode usually operates in reverse breakdown, it is often called an *avalanche diode*.

Zener Diode Operation

The operation of a zener diode is best understood through the use of an operating characteristic curve. See Figure 3-39. The overall forward and reverse characteristics of the zener diode are similar to the standard diode. When a source voltage is applied to the zener diode in the forward direction, there is a breakover voltage and forward current. When a source voltage is applied to the zener diode in the reverse direction, the current remains very low until the reverse voltage reaches reverse breakdown (zener breakdown). The zener diode then conducts heavily or avalanches. This is the important difference between zener diodes and standard diodes. When the zener diode conducts, it may continue to conduct for some time and at considerable current without damage in the reverse directions. This property is very important because the voltage drop across the zener diode remains essentially constant despite very large current fluctuations.

The zener diode is capable of being a constant voltage source because of the resistance changes that take place within the PN junction. This can be seen by studying the reverse operating characteristic curve. See Figure 3-40.

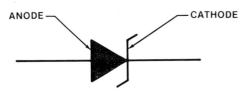

Figure 3-38. A zener diode symbol is different from a standard diode symbol in that the normally vertical cathode line is bent slightly at each end.

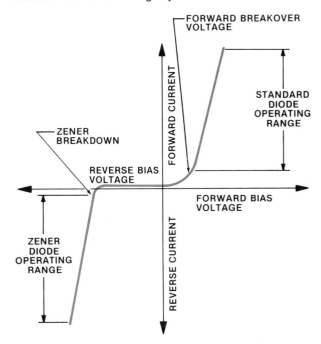

Figure 3-39. The overall forward and reverse characteristics of the zener diode are similar to the standard silicon diode. However, the zener diode normally operates in reverse breakdown and is often called an avalanche diode.

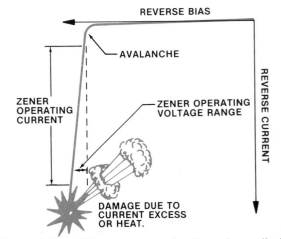

Figure 3-40. When a source of voltage is applied to the zener diode in the reverse direction, the resistance of the PN junction remains quite high and should produce only leakage current in the microampere (µA) range. As the reverse voltage is increased, the junction reaches a critical voltage and avalanche occurs.

When a source of voltage is applied to the zener diode in the reverse direction, the resistance of the PN junction remains high and should produce only leakage currents in the microampere range. However, as the reverse voltage is increased, the PN junction reaches a critical voltage and the zener diode avalanches. As the avalanche voltage is reached, the normally high resistance of the PN junction drops to a low value and the current increases rapidly. The current is limited generally by a circuit resistor or resistance R_L. See Figure 3-41.

This breakdown current should not be destructive to the zener diode. It becomes destructive if the current becomes excessive or the heat dissipating capabilities of the zener diode are exceeded.

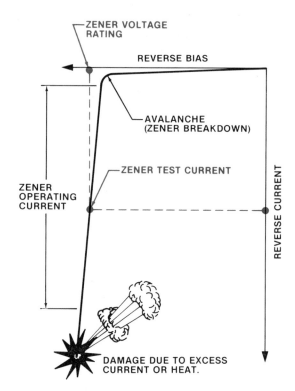

Figure 3-42. The rated voltage value of a zener diode is based on a specific zener test current. This value falls within the initial breakdown range and the maximum reverse current that a zener diode can safely handle.

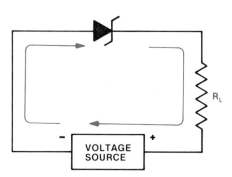

Figure 3-41. As the avalanche voltage is reached, the normally high resistance of the PN junction drops to a low value and the junction current increases rapidly. The current is generally limited by a circuit resistor or resistance R_L.

Zener Voltage Ratings

The zener diode is manufactured through a highly controlled doping process. The doping process allows the manufacturer to predetermine the operating range of a zener diode. Typical values of zener voltages may range from a few volts to several hundred volts.

When a zener diode is rated at a specific voltage, it is not necessarily the same value of voltage that begins to cause the diode to break down. The rated value represents the reverse voltage across the zener diode at a specific zener test current. This value falls within the initial breakdown range and the maximum reverse current that a zener diode can safely handle. See Figure 3-42.

Zener Tolerances

A zener diode is designed to have a specific breakdown voltage rating. Usually the voltage is a close approximation of the voltages necessary for circuit control, for example: 4.7, 5.1, 6.2, and 9.1 volts. Like resistors, zener diodes are manufactured according to certain tolerances. The standard zener breakdown voltage tolerances are ±20%, ±10%, or ±5%. When precision is required, zener diodes are also available at ±1%.

Obtaining Other Voltages

When a voltage that is not normally obtainable from standard zener diodes must be regulated, it is permissible to series connect several zener diodes to achieve proper voltage. See Figure 3-43. The zener diodes do not necessarily have equal breakdown voltages because the series arrangement is self-equalizing. However, their wattage ratings and current ranges should be similar, or the loads should be matched to avoid damaging any of the zener diodes.

Zener Wattage Rating

Zener diodes range from one millionth of a watt up to 50 watts. The most popular zener diodes are those rated at 1 watt and lower. The power dissipation rating

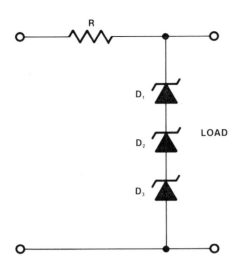

Figure 3-43. When it is necessary to regulate a voltage not normally obtainable from standard zener diodes, it is permissible to series connect several zener diodes to achieve the proper voltage.

of a zener diode, like other diodes, is given for a specific operating temperature. The rating is usually based on an ambient temperature of 25 °C.

Temperature Considerations

The zener breakdown voltage varies considerably with changes in ambient temperature. For this reason, the manufacturer frequently lists the zener voltage *temperature coefficient*. Temperature coefficient identifies the percentage of change in zener voltage per degree change in temperature. Typically, the zener voltage change is about 0.1% per degree centigrade.

ZENER DIODE TEST

A zener diode either provides voltage regulation, or it fails. If it fails, the zener diode must be replaced in order to return the circuit to proper operation. Occasionally, the zener diode may appear to fail only in certain situations. These types of failures are called *intermittents*. To check for intermittents, the zener diode must be tested while in operation. An oscilloscope is used for testing the characteristics of a zener diode in an operating situation. See Figure 3-44.

An oscilloscope displays the dynamic operating characteristics of the zener diode. If the zener diode is good, the appropriate test display will be shown. See Figure 3-45. The horizontal axis of the oscilloscope represents the voltage across the zener diode while the

vertical axis represents the current. This test is often used for production testing and sorting diodes.

ZENER DIODE APPLICATIONS

The zener diode is used in a variety of circuit applications. Because of its unique operating characteristics, it may be applied to very simple as well as very complex circuitry. However, the basic operating principle of the zener diode never changes.

Zener Diode as a Regulator

A zener diode used as a voltage regulator is usually installed in series with a resistor R_1, and in parallel with the load R_L. It is called a *series regulator*. See Figure 3-46. If it is installed in parallel, it is called a *shunt*

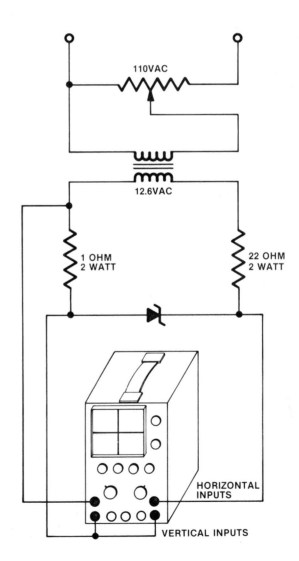

Figure 3-44. A zener diode that fails intermittently must be tested with an oscilloscope while it is operating in the circuit.

regulator. The total current through R_1 is the sum of the zener current and the load current.

If the voltage source E_S increases, the current through R_1, the zener diode, and R_L should increase. However, this does not happen because the zener diode acts as a variable resistor. When the voltage approaches the zener breakdown value, the resistance of the zener diode decreases. The reduced resistance causes the current to rise through the zener diode, causing a voltage drop across R_1. The voltage drop across R_1 lowers the output voltage and brings the voltage across R_L back to normal.

Decreasing voltage across the zener diode causes the current through it to decrease, and R_1 drops less voltage. This in turn raises the output voltage to normal.

Load changes also affect the zener diode as a regulator. As the load decreases or increases, the zener diode, acting as a shunt, will draw more or less current, respectively. The result of this change is a relatively constant output voltage across the regulator.

Operated within specifications, the zener voltage should not vary more than 0.1 volts. Zener diodes are designed to deliver constant voltage ratings ranging from a few volts to several hundred volts.

Zener Clipping and Limiting

A zener diode that is supplied with alternating current and connected as a shunt regulator is capable of limiting either the positive or negative parts of an AC cycle. See Figure 3-47, top. On the positive segment

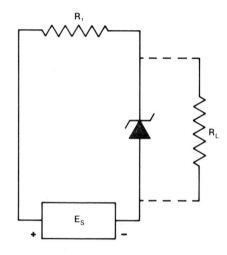

Figure 3-46. A zener diode used as a voltage regulator is usually installed in series with a resistor R_1 and in parallel with the load R_L. It is called a series regulator.

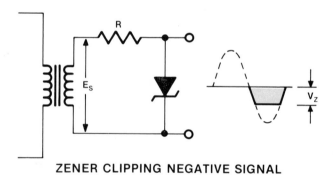

ZENER CLIPPING NEGATIVE SIGNAL

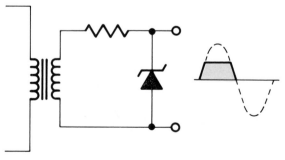

ZENER CLIPPING POSITIVE SIGNAL

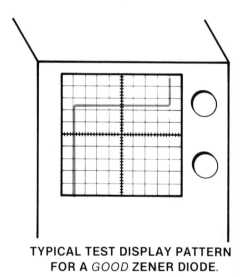

TYPICAL TEST DISPLAY PATTERN FOR A *GOOD* ZENER DIODE.

Figure 3-45. If a zener diode is good, the appropriate test display will appear. The horizontal axis of the oscilloscope represents voltage across the diode junction while the vertical axis represents the junction current.

Figure 3-47. A zener diode is capable of limiting either the positive or negative segments of an AC cycle when supplied with alternating current and connected as a shunt.

of the signal, the zener diode conducts almost immediately after the signal passes through zero, leaving no output signal across it. On the negative segment of the cycle, the zener diode does not conduct until the applied voltage reaches V_z. See Figure 3-47, top. At V_z the diode conducts, clipping the remainder of the signal (zener clipping). If the zener diode is reversed in the circuit, the positive signal will be clipped. See Figure 3-47, bottom.

If both the positive and negative portions of the signal are to be clipped, then the zener diodes must be installed back-to-back. See Figure 3-48. This type of circuit may be used to produce a relatively simple square wave signal. See Figure 3-49. It may also be used as a protection circuit. For example, the zener diodes could be wired across a speaker. See Figure 3-50. In this type of circuit, the zener diodes limit or clip any high-voltage spikes that could damage the speaker.

Simple Oscilloscope Calibrator

A stable voltage source, which is not affected by line voltage variations, is often necessary to calibrate an oscilloscope. A zener diode can be easily incorporated into existing oscilloscope circuitry to provide this calibration voltage. See Figure 3-51. A selected 10-volt zener diode is used to provide a calibration of one volt per division. Resistor R_L establishes the basic zener test current. Variable resistor R_S is used for fine adjustment. The output of this circuit is a square wave. However, the square wave is not perfect.

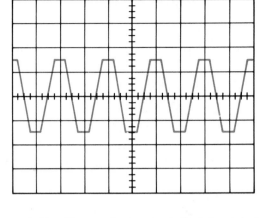

Figure 3-49. Zener diodes installed back-to-back produce a relatively simple square wave signal.

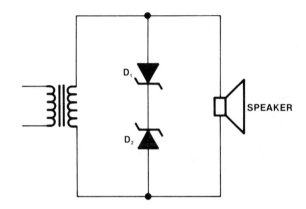

Figure 3-50. Zener diodes installed back-to-back may be used as a protection circuit for a speaker.

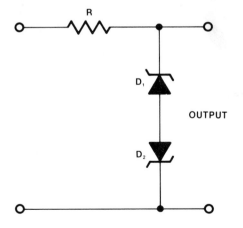

Figure 3-48. If both the positive and negative portions of the signal are to be clipped, the zener diodes may be installed back-to-back.

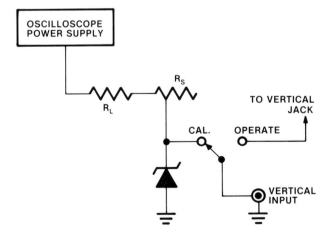

Figure 3-51. A zener diode can be easily incorporated into existing oscilloscope circuitry to provide a calibration voltage.

Chapter 3 - Review Questions

1. What makes diodes unique?
2. What is forward bias?
3. What is reverse bias?
4. How does current flow through a diode in reference to the anode and cathode?
5. What is the simplest form of a rectifier circuit?
6. What is a rectifier?
7. What is a semiconductor?
8. How is a semiconductor doped?
9. What is the difference between N-type material and P-type material?
10. What are carriers?
11. Describe an operating characteristic curve and its use.
12. What is peak inverse voltage (PIV)?
13. What type of tests can be made on a diode using an ohmmeter?
14. List the precautions that should be observed when using an ohmmeter.
15. What special precautions should be observed when installing diodes?
16. Define power capacity of a diode.
17. What is a zener diode?
18. How are zener diodes rated?
19. How are zener diodes tested?
20. What is the most common use for a zener diode?

4 DC POWER SUPPLIES— SINGLE PHASE

It is more practical to generate and transmit AC power than DC power. However, electronic circuits often require DC power to operate properly. When DC power is necessary, a *DC power supply* is used to produce the necessary voltages. The function of the DC power supply is to provide a regulated DC voltage.

A block diagram shows the basic construction of a DC power supply. See Figure 4-1. Each block indicates the components needed to make up most power supplies. The *transformer* is used to either step up or step down the AC line voltage to the operating voltage for the circuit. The diode *rectifier* changes the AC voltage to DC voltage. The *filter* section smooths out the output waveform of the rectifier. The *voltage regulator* maintains a relatively constant value of output over a wide range of operating situations. The *voltage divider* distributes the proper voltage to each load.

Depending upon cost and application, the DC power supply will contain some or all of the components shown in Figure 4-1.

Key Words

Capacitor Input Filter Circuit	L-Section Inductive Filter	Surge Current
DC Power Supply	L-Section Resistive Filter	Transformer
Filter	Pi-Section Filter	Unregulated Power Supply
Full-Wave Bridge Rectifier	Pulsating DC	Voltage Divider
Full-Wave Doubler	Rectifier	Voltage Doubler
Full-Wave Rectifier	Regulated Power Supply	Voltage Multiplier
Half-Wave Doubler	Ripple Voltage	Voltage Regulator
Half-Wave Rectifier		

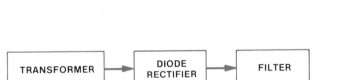

Figure 4-1. A block diagram shows the basic construction (components) of a DC power supply.

RECTIFIERS

Three basic types of rectifiers used in single-phase DC power supplies are the *half-wave rectifier*, *full-wave rectifier*, and *full-wave bridge rectifier*. See Figure 4-2.

Half-Wave Rectifiers

The operation of a half-wave rectifier is best understood by observing the effect the diode has on a typical AC input sine wave. See Figure 4-3. Half-wave rectification is accomplished because current is allowed to flow only when the anode terminal of the diode *D* is positive with respect to the cathode. Current is not allowed to flow through the rectifier when the cathode is positive with respect to the anode.

The output voltage of the half-wave rectifier is considered *pulsating DC* with half of the AC sine wave cut off. See Figure 4-3, right. The rectifier can pass either the positive or negative half of the input AC cycle, depending on the way the diode is connected into

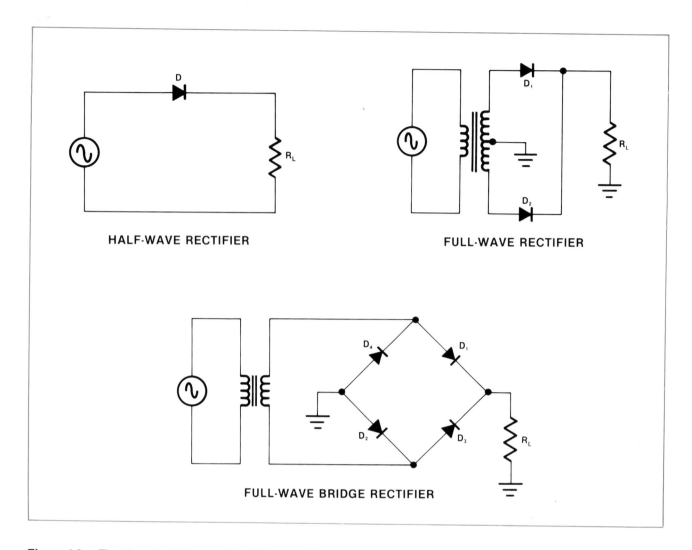

Figure 4-2. The three basic types of rectifiers used in single-phase DC power supplies are the half-wave, full-wave, and full-wave bridge.

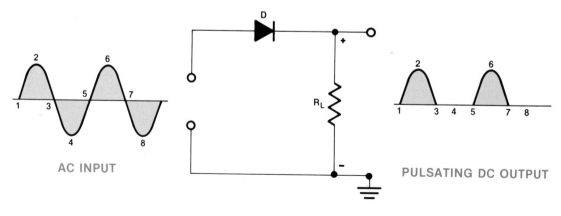

NOTE: THE NUMBERS ON THE SINE WAVE ARE FOR REFERENCE ONLY.

Figure 4-3. A half-wave rectifier is used to convert alternating current (AC) to pulsating direct current (DC).

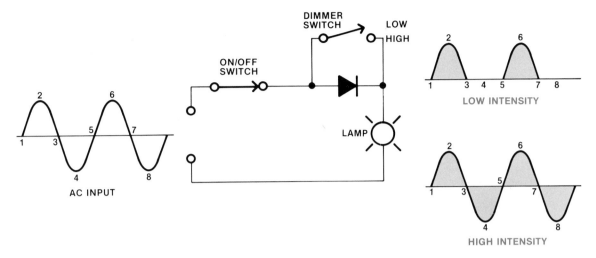

NOTE: THE NUMBERS ON THE SINE WAVE ARE FOR REFERENCE ONLY.

Figure 4-4. The half-wave rectifier is used in typical two-position light dimmers and some high-intensity lights to reduce the brightness.

the circuit. Half-wave rectification is considered inefficient for many applications because one-half of the input cycle is not used.

However, one application that makes use of the reduced current of the half-wave rectifier is a light dimmer. See Figure 4-4. This circuit is typical of two-position light dimmers and high-intensity lamps that are switched from bright to dim. When the dimmer switch is closed, or in the high position, the diode is shorted out, allowing full current and full brightness. When the dimmer switch is open, or in the low position, the diode is placed in series with the lamp, blocking one-half the current. This reduces the intensity of the lamp.

Full-Wave Rectifier

For full-wave rectification, an additional diode and a center-tapped transformer is utilized. See Figure 4-5. The center-tapped transformer supplies out-of-phase voltages to the two diodes.

When diode D_1 has a positive voltage applied, it is forward biased and conducts. See Figure 4-6, left. When D_1 conducts, current flows from the center-tapped transformer through the load resistor R_L and back to the transformer, resulting in a particular waveform. See Figure 4-6, right. The second diode D_2 shown in Figure 4-7, left, conducts under similar conditions, resulting in the waveform shown in Figure 4-7, right.

The output voltage of the full-wave rectifier has no off cycle. Electrons flow through the load during both half cycles. The constant current flow results in a complete output signal. See Figure 4-8. The full-wave rectifier is more efficient and has a smoother output than the half-wave rectifier.

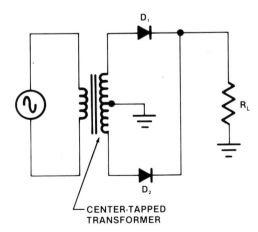

Figure 4-5. A full-wave rectifier utilizes an additional diode and a center-tapped transformer.

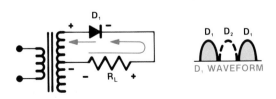

Figure 4-6. Diode D_1 with a positive voltage applied is forward biased. It conducts, resulting in the D_1 waveform (right).

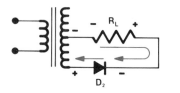

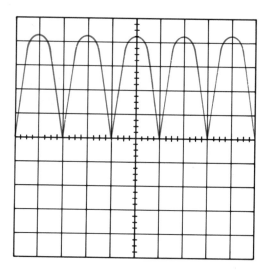

Figure 4-7. Diode D_2 with a positive voltage applied is forward biased. It conducts, resulting in the D_2 waveform (right).

Figure 4-8. The oscilloscope screen shows that the full-wave rectifier is more efficient and has a smoother output than the half-wave rectifier.

Full-Wave Bridge Recitifer

The full-wave bridge rectifier circuit consists of four diodes, but it does not require a center-tapped transformer. See Figure 4-9. It requires lower voltage diodes than the center-tapped circuit of the full-wave rectifier. The bridge diodes need to block only half as much reverse voltage as the center-tapped diodes for the same output voltage.

In Figure 4-9, top, the AC supply voltage is positive at point A and negative at point B. This allows current to flow from point B to point C, on through the load to point D, then to point A. Current, therefore, is passing through two diodes, D_1 and D_2. When the AC supply voltage is positive at point B and negative at point A, current flows from point A to point C, through the load to point D, then to point B. See Figure 4-9, bottom. This allows the current to pass through the other two diodes, D_3 and D_4.

One disadvantage of the full-wave bridge rectifier is that on each alternation, the direct current in the circuit must flow through two series-connected diodes. The forward DC voltage drop (loss) across the two rectifiers is therefore greater than the drop across a single rectifier. However, the voltage drop across silicon diodes is small (0.6 volts) and the loss can usually be tolerated.

AC AND DC VOLTAGE MEASUREMENTS IN RECTIFIER CIRCUITS

Measuring voltages in a rectifier circuit requires knowledge of some of the terminology, math, and instruments necessary to take accurate measurements. A rectifier circuit utilizes both AC and DC voltage. Care must be exercised when measuring, converting, or comparing any AC voltage value to a DC voltage value. For example, 100-volts AC is not necessarily the same as 100-volts DC. An AC voltage is always changing in value while a DC voltage is a continuous

value. Comparing one value to another may require a mathematical conversion, depending upon the test instrument used to take the measurement.

AC and DC Instruments

Measurements in a rectifier circuit are generally taken with either a voltmeter or an oscilloscope. See Figure 4-10. The DC voltmeter measures the average voltage. The AC voltmeter measures effective, or RMS (root mean square) voltage. The oscilloscope measures peak-to-peak (PP) voltage. A table is usually used to convert one value to another. See Figure 4-11.

If an oscilloscope is used, care must be exercised to properly identify the parts of the waveform so that the trace (scope pattern) is accurately converted for comparison purposes. See Figure 4-12. The major values to recognize on an AC sine wave are: peak, peak-to-peak, instantaneous, RMS, and average. See Figure 4-12.

Peak value is the maximum value of an alternating voltage that is reached on the positive or negative alternation. Peak-to-peak value is the maximum value measured from the negative peak to the positive peak. Instantaneous value is the value of voltage at any one instant. RMS value is equal to 0.707 times the peak value. Average value is the average of all the instantaneous values during one alternation, either positive or negative.

Mathematical Conversion of AC Values

To convert from RMS to PP or PP to average, refer to the conversion table of Figure 4-11. All conversions are done with peak values, not peak-to-peak values.

For example, in converting 311V PP to RMS, 311V PP must first be divided in half to obtain a value of 155.5V Peak (Pk). The 155.5V peak value is then multiplied by 0.707 (155.5 × 0.707) to obtain 109.94V RMS, or approximately 110V RMS. See Figure 4-13.

Voltage Output of a Half-Wave Rectifier

Once the AC input voltage of a rectifier is determined, the output voltage can be found by taking a direct reading with a DC voltmeter. Using an oscilloscope and making some calculations is another method used

to find the output voltage. See Figure 4-14. If an oscilloscope is used, connect it across the output. The value of the AC output will be one-half the peak-to-peak value of the input minus the 0.6 volt drop across the diode. To convert from peak value to average DC value, use the conversion table of Figure 4-11. In this circuit, 155.5 volts minus the 0.6 volts across the diode, or 154.9 volts × .637 gives an average value of 98.67. Since every other pulse is missing, the average value will be reduced to 49.34 volts (the average DC value).

NOTE: THE NUMBERS ON THE SINE WAVE ARE FOR REFERENCE ONLY.

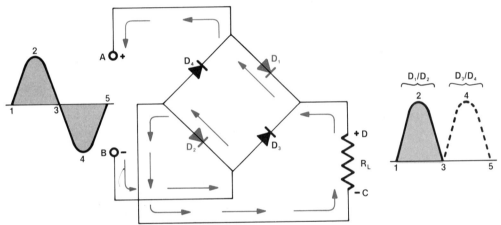

D₁ AND D₂ ARE FORWARD BIASED AND CONDUCTING WHILE D₃ AND D₄ ARE REVERSED BIASED AND ARE NOT CONDUCTING.

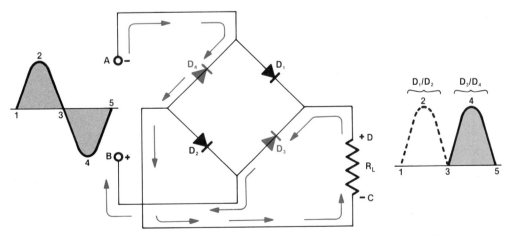

D₃ AND D₄ ARE FORWARD BIASED AND CONDUCTING WHILE D₁ AND D₂ ARE REVERSED BIASED AND ARE NOT CONDUCTING.

Figure 4-9. The full-wave bridge rectifier requires four doides, but it does not require a center-tapped transformer.

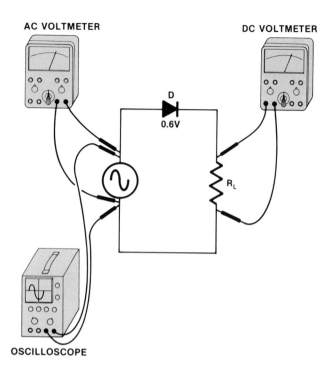

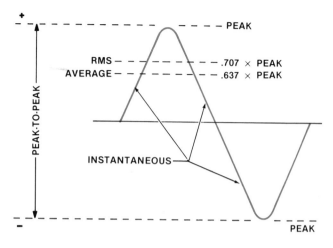

Figure 4-12. The major values to recognize on an AC sine wave are: peak, peak-to-peak, instantaneous, RMS, and average.

Figure 4-10. A voltmeter or an oscilloscope is usually used to take measurements in a rectifier circuit.

GIVEN VALUE	TO SOLVE FOR		
	PEAK	RMS	AVERAGE
PEAK	—	.707 × PK	0.637 × PK
RMS	1.414 × RMS	—	.9 × RMS
AVERAGE	1.57 × AVG	1.11 × AVG	—

Figure 4-11. A table is often used to compare the relationship between peak, RMS, and average voltage readings in a circuit.

Voltage Approximations

When troubleshooting a rectifier circuit, precise voltage measurements are rarely necessary. Initially, it is only necessary to determine whether or not the device is operating. Approximated voltages give enough information on the condition of the circuit to make intelligent decisions.

To determine the expected average DC voltage of a full-wave rectifier, multiply the actual effective voltage (measured at the rectifier input) by 90%. This factor (90%) has been predetermined by dividing average voltage by effective voltage:

$$\frac{63.2}{70.7} = .90$$

In a half-wave rectifier circuit, one-half of the input sine wave is removed by the rectifier. Therefore, the average DC voltage should be 45% of the effective voltage:

$$\text{Average DC Voltage} = \frac{\text{RMS value} \times .90}{2}$$

$$= \text{RMS} \times .45$$

For the half-wave rectifier circuit of Figure 4-14, the average DC voltage is:

$$110 \times .45 = 49.5\text{V}$$

Since one diode is in that circuit, a voltage drop of 0.6 should be subtracted (49.5 − 0.6) for a net result of 48.90 volts. This application is close enough to the calculated value of 49.34 volts and therefore serves its purpose.

Voltage Output of a Full-Wave Rectifier

Since the full-wave rectifier uses both alternations of an AC cycle, it supplies twice the amount of DC voltage as a half-wave rectifier. Instead of 45% of the effective value (RMS value), the result at the load will be 90% of the effective AC value. To determine the

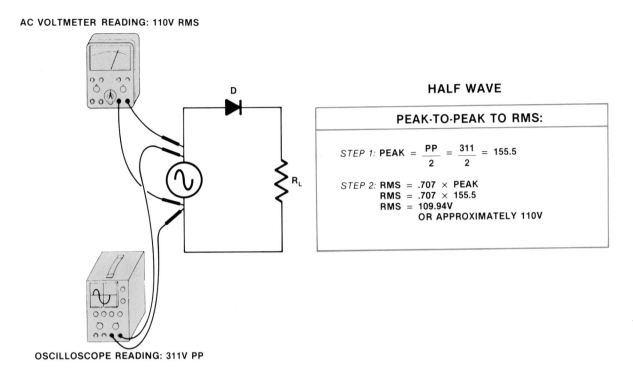

AC VOLTMETER READING: 110V RMS

HALF WAVE

PEAK-TO-PEAK TO RMS:

STEP 1: PEAK $= \dfrac{PP}{2} = \dfrac{311}{2} = 155.5$

STEP 2: RMS $= .707 \times$ PEAK
RMS $= .707 \times 155.5$
RMS $= 109.94$V
OR APPROXIMATELY 110V

OSCILLOSCOPE READING: 311V PP

Figure 4-13. Conversions from one value to another must always be made in relation to the peak value.

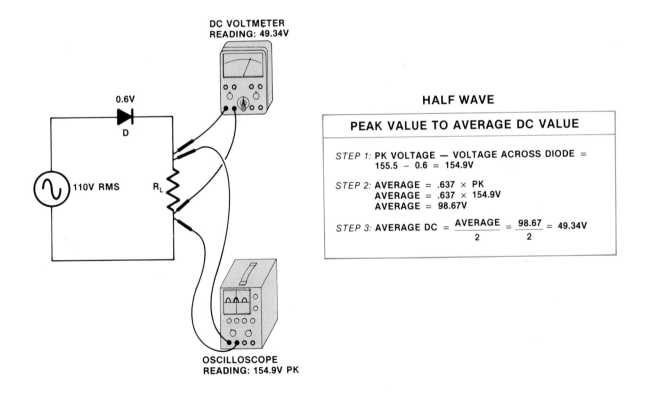

DC VOLTMETER
READING: 49.34V

HALF WAVE

PEAK VALUE TO AVERAGE DC VALUE

STEP 1: PK VOLTAGE — VOLTAGE ACROSS DIODE =
155.5 − 0.6 = 154.9V

STEP 2: AVERAGE $= .637 \times$ PK
AVERAGE $= .637 \times 154.9$V
AVERAGE $= 98.67$V

STEP 3: AVERAGE DC $= \dfrac{\text{AVERAGE}}{2} = \dfrac{98.67}{2} = 49.34$V

OSCILLOSCOPE
READING: 154.9V PK

Figure 4-14. The output of a half-wave rectifier can be measured by taking a direct reading with a DC voltmeter. It can also be found by using an oscilloscope and making calculations with conversion tables.

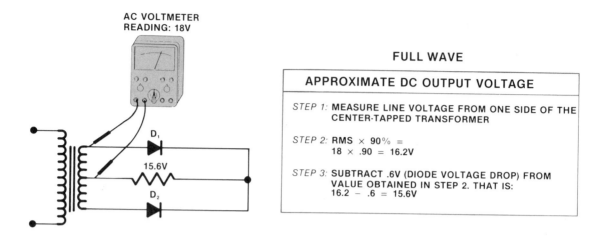

Figure 4-15. The output of a full-wave rectifier should be approximately 90% of the effective (RMS) AC input minus the 0.6 volt drop across the diode.

approximate output voltage of the circuit in Figure 4-15, measure the actual line voltage from one side of the transformer to the center tap. Multiply this value (18 volts) by 90% to obtain 16.2 volts. Subtract the 0.6 volts from the voltage drop of diode D_1 to obtain a final value of 15.6 volts. The value of only one diode is subtracted because only one diode will pass current at any given time.

Voltage Output of a Full-Wave Bridge Rectifier

Like the output from a full-wave rectifier, the output of the full-wave bridge rectifier is 90% of the effective AC input. However, current in a bridge rectifier is always flowing through two diodes. Therefore, when calculating the expected DC output of a full-wave bridge rectifier, the voltage drop for two diodes must be subtracted.

For instance, to determine the approximate output voltage in Figure 4-16, measure the actual line voltage across the bridge circuit. Multiply this value (50 volts) by 90% and obtain a value of 45 volts. Subtract 1.2 volts (the voltage drop of the two diodes) from the 45 volts to obtain a final value of 43.8 volts.

TESTING HALF-WAVE RECTIFIERS

Although half-wave rectifiers have relatively few parts, the following problems may arise:

1. A fuse or circuit breaker may be open in the primary or secondary of the circuit causing a loss of voltage.

2. The transformer could be shorted or open; no voltage would appear across the circuit.

3. An open circuit in the diode will be indicated by a voltmeter reading of 0 volts at the load.

4. A short circuit in the diode will be indicated by current passing in both directions and full AC voltage across the load.

TESTING FULL-WAVE RECTIFIERS

Problems that may arise in full-wave rectifiers are:

1. A fuse or circuit breaker may be open in the primary or secondary causing a loss of voltage.

2. The transformer could be shorted or open; no voltage would appear across the circuit.

3. If diode 1 or diode 2 were open or shorted, the output voltage would drop to one-half voltage.

4. If both diodes were open, there would be zero volts at the load.

5. If both diodes were shorted, the entire transformer secondary would be shorted out.

TESTING FULL-WAVE BRIDGE RECTIFIERS

Problems that may arise in full-wave bridge rectifiers are:

1. A fuse or circuit breaker may be open in the primary or secondary causing a loss of voltage.

2. The transformer could be shorted or open; no voltage would appear across the circuit.

3. If one of the diodes opens, the DC output will equal that of a half-wave rectifier, minus the voltage drop of the two diodes.

4. If one of the diodes is shorted, the voltage output will be similar to a half-wave rectifier.

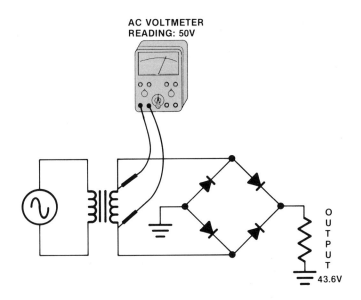

FULL-WAVE BRIDGE

APPROXIMATE DC OUTPUT VOLTAGE
STEP 1: MEASURE LINE VOLTAGE ACROSS THE BRIDGE CIRCUIT.
STEP 2: RMS × 90% = 50 × .90 = 45V
STEP 3: SUBTRACT 1.2 VOLTS (VOLTAGE DROP OF TWO DIODES) FROM VALUE OBTAINED IN STEP 2. THAT IS: 45 − 1.2 = 43.8V

Figure 4-16. The output of a full-wave bridge rectifier should be approximately 90% of the effective (RMS) AC input minus the 1.2 volt drop across two diodes.

5. Another check with a voltmeter can be made across the diode. If the diode is shorted, the voltage will be zero.

When replacing diodes, the PIV is critical. For a full-wave bridge rectifier, the PIV is divided across the rectifiers. Hence, the PIV for each rectifier is the transformer peak. In the conventional two-diode full-wave rectifier, the PIV is approximately twice the transformer voltage.

POWER SUPPLY FILTERS

The filter section of a power supply smooths the pulsating DC to make it more consistent. A filter minimizes or removes ripple voltage from a rectified output by opposing changes in voltage and current.

The filtering process is accomplished by connecting parallel capacitors and series resistors or inductors to the output of the rectifier. The capacitors smooth the voltage while the inductors (chokes) smooth the current.

The capacitance of a capacitor determines its effectiveness in filtering voltages. With a larger capacitance, the capacitor filters more effectively. The filtering action of a choke is also directly related to its electrical size (inductance).

Capacitive Filter

The capacitive filter is a capacitor connected in parallel with the load resistance. See Figure 4-17. As the pulsating DC voltage from the rectifier is applied across capacitor C, it charges to the peak applied

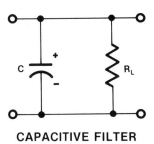

CAPACITIVE FILTER

Figure 4-17. The capacitive filter is a capacitor C connected in parallel with a load resistance R_L.

voltage. Between peaks, the capacitor discharges through the load resistance R_L and the voltage gradually drops. See Figure 4-18. The amount of voltage drop before the capacitor begins to charge again is called the *ripple voltage*. See Figure 4-19. The amount of capacitor discharge between voltage peaks is controlled by the RC time constant of the capacitor and the load resistance. If the load resistance is large and the capacitance is large, the ripple voltage will be small, resulting in a smooth output.

The first filter element in the capacitive filter is a capacitor. It is called a *capacitor input filter circuit*. The capacitor input filter circuit provides maximum voltage output to the load. Since a large capacitor is needed, an electrolytic capacitor is usually connected with the polarity. See Figure 4-18.

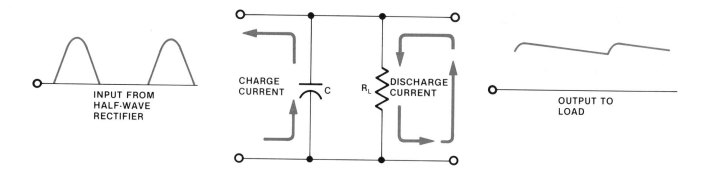

Figure 4-18. As the pulsating DC voltage from the rectifier is applied across the capacitor C, it charges to the peak applied voltage. Between peaks, C discharges through the load resistance R_L and the voltage gradually drops.

L-Section Resistive Filter

An *L-section resistive filter* reduces surge currents by using a current-limiting resistor R_1. See Figure 4-20. R_1 controls surge currents by using an *RC* time constant to slow the charging of the capacitor. R_1 should always be used in series with the rectifier and the input capacitor of the filter system. This protects the rectifier from the high surge of charging current that flows through the rectifier from the input capacitor when the circuit is first energized. A resistor of about 50 ohms (Ω) is usually used in this application. The filtering action of the resistor is not as good as that of an inductor, but it is much cheaper.

L-Section Inductive Filter

An *L-section inductive filter* reduces surge currents by using a current-limiting inductor L_1. See Figure 4-21. An inductor used as the series element opposes a change in current by creating a counter electromotive force (CEMF), or counter voltage. The result is that the surge current is greatly reduced and the capacitor charges slowly. The inductor also aids the filtering action of the capacitor since the CEMF of the inductor tends to cancel out the effects of the ripple voltage.

The operation of the L-section inductive filter can also be seen through the effect that inductive reactance has on the circuit. When the pulsating DC voltage is applied to the inductor, the changing voltage produces a high inductive reactance. Therefore, the inductor tends to block the pulsating DC voltage. The DC portion of the signal does not create inductive reactance and is allowed to pass through the inductor readily. The pulses not blocked by the inductor are bypassed by the capacitor.

Disadvantages of Capacitive Filters. Ripple voltage increases when the load increases on the capacitive filter. Capacitive filters also tend to produce excessive currents. The excessive current, called *surge current*, flows into a capacitor when it is first turned on.

Over a period of time, surge currents may damage fuses and diodes in the circuit. Other power supply filters have components that reduce the effect of ripple-voltage variations and surge current.

Pi-Section Filter

A *pi-section filter* gets its name from the Greek letter pi because the filter configuration resembles the letter's symbol (π). The two types of pi filters are inductive and resistive.

A pi-section filter has three elements. In a pi-section inductive filter, there is a shunt input capacitor C_1, a series inductor L_1 (choke), and a shunt output

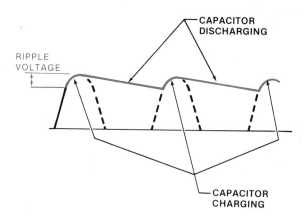

Figure 4-19. The amount of voltage drop before the capacitor begins to charge again is called ripple voltage.

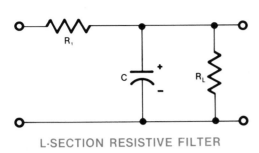

L-SECTION RESISTIVE FILTER

Figure 4-20. The L-section resistive filter reduces surge currents by using a current-limiting resistor R_1.

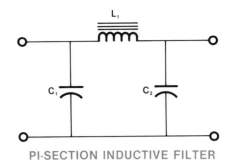

PI-SECTION INDUCTIVE FILTER

Figure 4-22. By using a shunt input capacitor C_1, a series inductor L_1, and a shunt output capacitor C_2, a relatively smooth DC voltage output is obtained from this pi-section inductive filter.

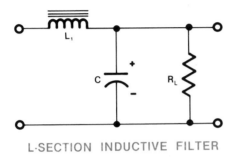

L-SECTION INDUCTIVE FILTER

Figure 4-21. The L-section inductive filter reduces surge currents by using a current-limiting inductor L_1.

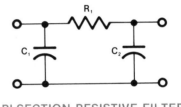

PI-SECTION RESISTIVE FILTER

Figure 4-23. When money is a factor, the inductor is sometimes replaced by a resistor in a pi-section filter. Resistive filters are not as effective as inductive filters.

capacitor C_2. See Figure 4-22. As the input voltage reaches the first capacitor C_1, the capacitor shunts most of the AC ripple current to ground. This presents a much smoother current to L_1. Since L_1 presents a high inductive reactance to the remaining AC ripple, L_1 tends to block the AC ripple. Finally, C_2 shunts to ground any remaining AC ripple. The result is a relatively smooth DC voltage.

When money is a factor, L_1 is sometimes replaced by a resistor R_1. See Figure 4-23. A pi-section filter that uses a resistor depends upon the use of the RC time constant. Although resistive filters accomplish filtering, they are not as effective as inductors.

FILTERS AND PEAK INVERSE VOLTAGE (PIV)

The PIV or peak inverse voltage of a diode should not be exceeded. The value of the PIV will be 1.414 times the RMS value of the secondary voltage of the transformer. However, when a filter is added to the circuit, the PIV becomes much higher. Figure 4-24 shows a half-wave rectifier with a pi-section inductive filter added.

In this circuit, when diode D conducts, capacitor C_1 charges to almost the peak value of the AC applied. C_1 holds its charge for some time. Therefore, its voltage adds to that voltage already across the diode when the sine wave begins its reverse alternation.

C_1 voltage adds to the voltage across the transformer. C_1 voltage increases the inverse voltage across the diode to almost twice the amount of the peak value of AC secondary voltage.

In a full-wave rectifier with a center-tapped secondary, the PIV is twice the peak value of one-half of the secondary voltage.

It may appear that the diode is rated much higher than necessary and that a lower-rated diode would appear to work. The circuit designer, however, has taken into account the additional PIV of the filter circuit. Therefore, a diode with a PIV rating equal to or higher than the one removed must always be used.

Increasing PIV and Current Ratings

Silicon rectifiers are connected in series for higher PIV ratings, and in parallel for higher current ratings. Figure 4-25 illustrates diodes connected in series to increase the PIV rating. In this case, if each diode were rated for 100 volts individually, they would be 200 volts in series. However, when silicon diodes are in series, each one develops 0.6 volts and must be added together as a voltage drop in the circuit. For example, two diodes in series have a combined voltage drop of 1.2 volts. To increase current rating, the diodes would be placed in parallel.

VOLTAGE REGULATION

Power supplies are either regulated or unregulated. It depends upon whether or not the final output voltage must be constant or if it can fluctuate over certain limits. A power supply whose output varies, depending upon changes of line voltage or load, is called an *unregulated power supply*. A power supply that maintains a constant voltage across the output under various conditions of line voltage load change is called a *regulated power supply*.

Many electronic devices require regulated power supplies. These include motor controls, computers, and critical timing equipment.

Series Regulator

The voltage regulator is a variable resistance that automatically changes as the output voltage changes. The output voltage of a rectifier is held constant by a variable resistor R_1 in series with the load R_L. See Figure 4-26. The load current flows through the variable resistor and causes a voltage drop across it. Thus, the variable resistor and the load develop a DC voltage divider. (See VOLTAGE DIVIDER.) The resistance to maintain a desired voltage across the load can be adjusted even when variations occur in the input voltage to the rectifier or when the load current changes.

As the load current increases, a larger voltage drop appears across the resistance. The voltage across R_L is then reduced. To maintain voltage across R_L, the value of resistance can be reduced. The reduction lowers the voltage across R_1 until the voltage across R_L returns to its required value.

NOTE: The limitation of this circuit is that it must be manually operated.

Shunt Regulator

A shunt regulator combined with the resistance of the power supply itself, or with an additional resistor, forms a voltage divider. See Figure 4-27. As the shunt resistance R_2 increases, more voltage appears across it as an output to the load. As R_2 decreases, less voltage appears across it.

The resistance of a shunt regulator increases when the output voltage decreases. The resistance of a shunt regulator decreases when the output voltage increases. Thus, the shunt regulator returns the output voltage to normal. (The zener diode is a type of shunt regulator.)

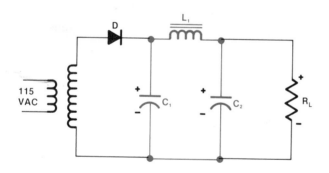

Figure 4-24. When a pi-section inductive filter is added to a rectifier circuit, the PIV of the diode must be increased due to the voltage being held by the capacitor C_1 on the reverse alternation.

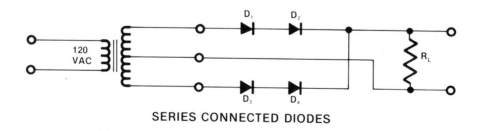

SERIES CONNECTED DIODES

Figure 4-25. Silicon rectifiers can be connected in series for higher voltage ratings or in parallel for higher current ratings. In this circuit, the diodes are in series to increase the voltage rating.

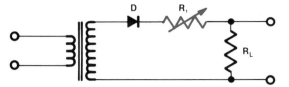

NOTE: VARIABLE RESISTOR R₁ ACTS AS THE VOLTAGE
REGULATOR.

Figure 4-26. The voltage regulator of a series-regulated power supply is a variable resistor in series that can change as the output voltage changes.

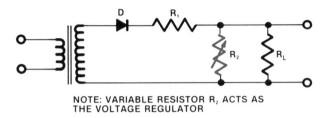

NOTE: VARIABLE RESISTOR R₂ ACTS AS
THE VOLTAGE REGULATOR

Figure 4-27. The voltage regulator of a shunt-regulated power supply is a variable resistor in parallel. It automatically changes as the output voltage changes.

VOLTAGE DIVIDER

A *voltage divider* is used in a power supply when it is necessary to provide voltages for several different loads. See Figure 4-28. It distributes the available voltage into those voltages required by the various loads.

The voltage divider is a resistor, or a series of resistors, connected across the output of a power supply. Sometimes it is tapped at different points to provide a selection of output voltages. The resistor, or series of resistors, are in parallel with the various loads. The total resistance of the voltage divider provides a "bleeder network." The total resistance appears to the power supply as a fixed load and a discharge path for the capacitors when the power supply has been turned off. This fixed load provides a degree of voltage regulation.

VOLTAGE MULTIPLIER

Voltage multipliers are circuits designed to supply high voltages with a limited load current. The advantage of the voltage multiplier is its ability to develop voltages higher than the AC source without using a step-up transformer. Thus, the voltage multiplier is often called a transformerless power supply. It is

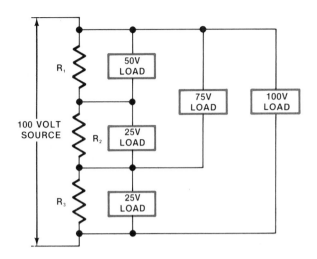

Figure 4-28. A voltage divider distributes the available voltage into those voltages required by the various loads.

capable of delivering a DC voltage as high as 400 volts from a 110-VAC source.

Figure 4-29 illustrates the basic operating principle behind the voltage multiplier. In this circuit, two capacitors are charged individually from a 100-VDC source. When the leads from the 100-VDC source are connected across C_1, it charges to 100 VDC with point A positive with respect to point B. See Figure 4-29, left. When the leads are moved to C_2 with the positive lead at point B and the negative lead at point C, C_2 also charges to 100 VDC. See Figure 4-29, right.

After both capacitors are charged, the 100-VDC source is disconnected. When a voltmeter is connected across the two capacitors, the reading is 200 VDC. See Figure 4-30. In series, the voltage across two capacitors equals the sum of the two individual voltages—in this case, 200 VDC.

Voltage multiplier circuits operate on the same principle. They are series-connected capacitors. However, rectifiers and resistors are used to produce and control the peak DC voltages.

Effect of RC Time on DC Output

The resistance in a half-wave transformerless rectifier circuit can be varied to control the value of DC output. See Figure 4-31. On the positive half cycle of input voltage, when point A is positive with respect to point B, diode D conducts and there is current through variable resistor R. With 110 VAC at the input, capacitor C charges to the peak value of 155.5 volts (110×1.414). On the next alternation, the anode of D is negative and it will not conduct. See Figure 4-32. During this time, C discharges through R. This action

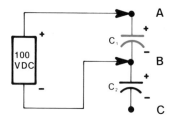
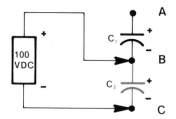

Figure 4-29. When the 100-VDC source is across C_1, the capacitor charges up to 100 VDC. When the 100-VDC source is across C_2, it also charges up to 100 VDC.

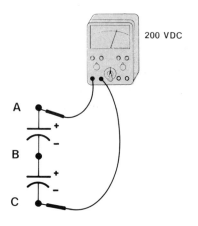

Figure 4-30. A voltmeter connected across both charged capacitors gives a reading of 200 VDC.

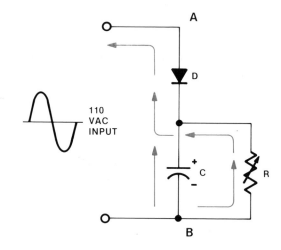

Figure 4-31. On the positive half cycle of the input voltage, diode *D* conducts and capacitor *D* charges to the peak value of the AC input.

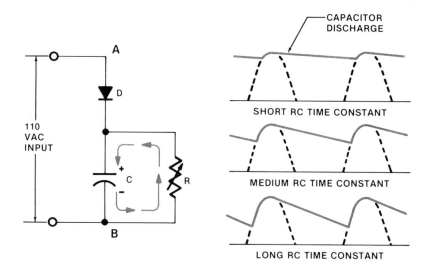

Figure 4-32. A larger average DC output voltage can be obtained by using a diode and an *RC* circuit than by using a simple resistive load.

repeats with each half cycle of input voltage. The result is a varying value of direct current across R.

Each time *D* conducts, *C* charges to the peak value of applied voltage. However, the extent to which it discharges during the nonconducting alternation depends upon the *RC* time of the resistance-capacitance combination. Therefore, an average direct output voltage is obtained. This voltage is of greater value than one obtained by using a simple resistive load. The variable discharge rate of this circuit for values of *R* and *C* can be illustrated to indicate the advantage. See Figure 4-32.

Voltage Doublers

Voltage Doublers are circuits designed to produce a DC output level that is approximately twice that of the peak AC input value. The two most common types of voltage doublers are the *half-wave doubler* and the *full-wave doubler*.

Half-Wave Doubler. The schematic diagram of a half-wave doubler circuit is shown in Figure 4-33. The circuit consists of two diodes, D_1 and D_2, two capacitors, C_1 and C_2, and a load resistance, R_L. The input is an AC voltage. During the positive half cycle of the AC input, C_1 charges because D_1 is forward biased. See Figure 4-34.

When the input signal is negative, C_1 is placed in series with the peak voltage of the supply signal. The

voltage stored in C_1 and the voltage source are combined. Therefore, the peak voltage applied to the circuit is doubled. Since D_2 is forward biased, C_2 is charged to this doubled peak voltage. See Figure 4-35. C_2 then discharges through R_L at twice the input voltage.

NOTE: There is current through R_L only during the half cycle when D_2 is conducting. When D_1 is conducting, C_1 charges to the peak value of the applied voltage and effectively doubles it for the duration of the time that D_2 conducts.

The disadvantage of this circuit is that the DC output contains a fairly large 60-hertz ripple. In addition, C_2 must be capable of operating at the full peak output voltage. This voltage is approximately three times the RMS value of the AC source under no-load conditions. The higher voltage rating raises the cost of the capacitor.

Full-Wave Doubler. A full-wave doubler uses both halves of the AC input signal. See Figure 4-36. Both capacitors are charged on alternate cycles of the input

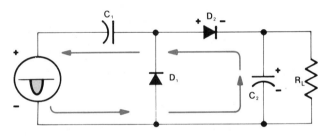

Figure 4-35. When the input signal is negative, diode D_2 is forward biased and capacitor C_2 charges. C_2 charges to double the peak voltage since C_1 and peak voltage of the supply are in series with C_2.

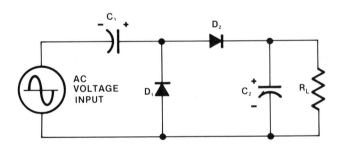

Figure 4-33. This half-wave doubler circuit consists of two diodes, two capacitors, and a load resistance.

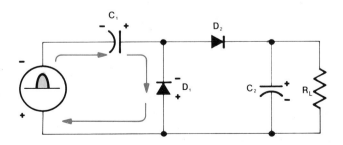

Figure 4-34. During the positive half cycle of the AC input, capacitor C_1 charges because diode D_1 is forward biased.

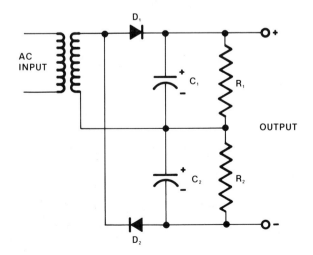

Figure 4-36. The full-wave doubler uses both halves of the input signal by charging two capacitors on alternate cycles.

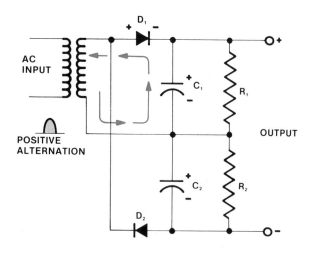

Figure 4-37. When the AC input is positive, diode D_1 conducts and charges capacitor C_1 to the peak value of the input voltage.

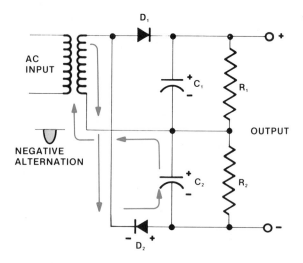

Figure 4-38. When the AC input is negative, diode D_2 conducts and charges capacitor C_2. Since C_1 and C_2 are in series across the output, the voltages combine.

so that both capacitor voltages are added together in the output.

Figure 4-37 illustrates one-half of the operation of a full-wave doubler. When the AC input is positive, diode D_1 conducts and charges capacitor C_1 to the peak value of the input with the polarity indicated. When the AC input is negative, diode D_2 conducts and

charges capacitor C_2 to the peak value of the input with a certain polarity. See Figure 4-38. Because C_1 and C_2 are in series across the output, the voltages combine. The regulation of the full-wave doubler is better than the half-wave doubler because one capacitor is always being charged in any given half cycle.

Chapter 4 - Review Questions

1. What is the main function of a DC power supply?
2. What are the major parts of most DC power supplies?
3. List the three basic types of rectifiers used in single-phase DC power supplies.
4. How much voltage is usually dropped across a silicon diode?
5. What type of voltage will the following instruments read?
   DC voltmeter
   AC voltmeter
   Oscilloscope
   Multimeter
6. How are AC and DC voltages converted from one value to another?
7. How is the average DC voltage of a full-wave rectifier approximated?
8. How is the average DC voltage of a half-wave rectifier approximated?
9. How is the average DC voltage of a full-wave bridge rectifier approximated?
10. List four ways the output of *any* rectifier can be affected.
11. What is the function of the filter section in a DC power supply?
12. What are the four types of filter?
13. What are the disadvantages of capacitive filters?
14. What affect does a filter have on PIV?
15. What are the three sections (elements) of a pi-section filter?
16. What are the two types of regulators and how do they work?
17. How can the PIV and current rating of a diode be increased?
18. What is a voltage divider and when is it used?
19. What are voltage multipliers and how do they work?
20. What effect does the RC time constant have on DC output?

5 SOLID STATE TRANSDUCERS

A *transducer* is any device that changes one form of energy to another form of energy. Transducers provide a valuable link in industrial control systems. They convert mechanical, magnetic, thermal, electrical, optical, and chemical variations into electrical voltages and currents. These voltages and currents are used directly or indirectly to drive other control systems. Because of the variety of solid state devices available, many types of electromechanical transducers are being replaced with solid state transducers.

Key Words

Alphanumeric display	Negative temperature coefficient	Pressure sensor
Cold resistance	Nematic liquid	Seven-segment display
Hall effect sensor	Photoconductive cell	Status indicator
Hall generator	Photoconductive diode	Switch temperature
Hot resistance	Photodiode	Temperature compensation
Infrared light	Photovoltaic cell	Thermistor
Light emitting diode	Polarizing filter	Transducer
Liquid crystal display	Positive temperature coefficient	

THERMISTOR

A *thermistor* is a thermally sensitive resistor. The resistance of a thermistor changes with a change of temperature. See Figure 5-1. As a match is placed under the thermistor, its resistance decreases and current flow increases. When the match is removed, the resistance increases to its original state (resistance value).

The operation of a thermistor is based on the electron-hole pair theory. As the temperature of the semiconductor increases, the generation of electron-hole pairs increases due to thermal agitation. Increased electron-hole pairs causes a drop in resistance.

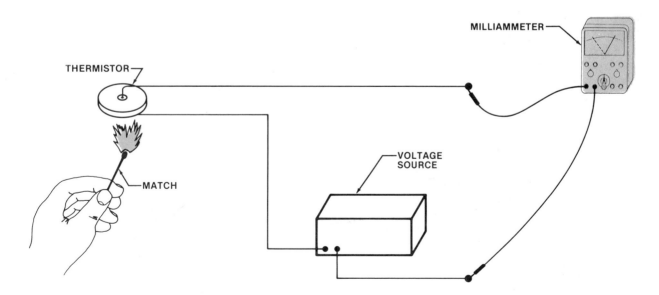

Figure 5-1. A thermistor is a transducer that acts as a thermally sensitive resistor. Its resistance changes with a change in temperature. An increase in temperature causes a decrease in resistance and an increase in current.

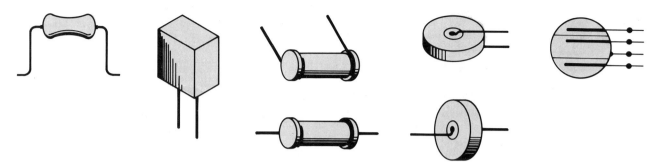

Figure 5-2. Thermistors are available in a variety of sizes and styles.

Thermistor Types

Figure 5-2 shows several types of thermistors. Thermistors are popular because of their small size. They can be mounted in places that are inaccessible to other temperature-sensing devices. For example, a thermistor can be embedded in the coil of a motor to provide additional thermal protection. Figure 5-3 shows the small size in which a thermistor can be made. Thermistors can also be made in the shapes of beads, disks, washers, and rods. The sizes and shapes vary depending upon their application.

Thermistor Resistance Values

The resistance value of a thermistor varies from a few ohms (Ω) to the megaohm ($M\Omega$) range. Thermistors are linear and nonlinear. In a linear thermistor, the resistance changes the same amount for each degree of temperature change. In a nonlinear thermistor, the resistance varies dramatically in different temperature ranges.

Thermistors are sensitive and can provide fractional degree temperature control. Some thermistors can be accurate to $\pm 0.1\,°C$ change in temperature. Since thermistors are resistive devices, they can also operate on AC or DC.

Thermistor Schematic Symbols

Figure 5-4 shows the schematic symbols for a directly heated and an indirectly heated (externally heated) thermistor. Directly heated thermistors are used in voltage regulators, vacuum gauges, and electronic time-delay circuits. Indirectly heated thermistors are used for precision temperature measurement and temperature compensation.

PTC and NTC

The two classes of thermistors are *positive temperature coefficient (PTC)* and *negative temperature coefficient (NTC)*. Although most thermistors operate on an NTC, some applications require operation on a PTC.

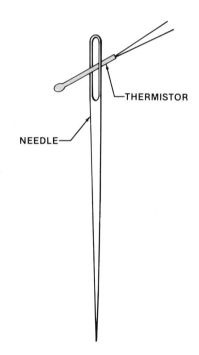

Figure 5-3. Thermistors are manufactured in small sizes that allow them to be mounted in places that are inaccessible to other temperature-sensing devices.

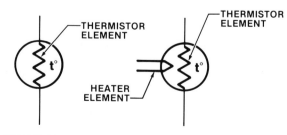

DIRECTLY HEATED   INDIRECTLY HEATED

Figure 5-4. The schematic symbols show that a thermistor is either directly or indirectly heated.

With a PTC, an increase in temperature causes the resistance of the thermistor to increase. With the NTC, an increase in temperature causes the resistance of the thermistor to decrease.

Cold and Hot Resistance

Cold resistance and *hot resistance* refer to the operating resistance of a thermistor at extremes of temperature. Cold resistance is measured at 25 °C, or room temperature. However, some manufacturers specify lower temperatures. The specification sheet should always be checked. Hot resistance is the resistance of a heated thermistor. In a directly heated thermistor, the heat is generated from the ambient temperature and the current passing through the device. In an indirectly heated thermistor, the heat is generated from the ambient temperature, the current, and the heater element of the thermistor.

NTC THERMISTOR APPLICATIONS

The resistance temperature characteristic curve for an NTC is shown in Figure 5-5. The large swing of resistance in relation to temperature enables the thermistor to become a versatile solid state device. The resistance temperature plot for an NTC thermistor is approximately exponential (nonlinear). Formulas and tables that can be used to make specific calculations are available from manufacturers.

NTC Thermistor for Fire Protection

A fire alarm circuit is typical of a circuit that requires an NTC thermistor. See Figure 5-6. The purpose of this circuit is to detect a fire and sound the alarm.

In normal operating environments, the resistance of the thermistor is high because ambient temperatures are relatively low. The high resistance keeps the current to the control circuit low. The alarm remains OFF. However, in the presence of a fire, the increased temperature lowers the resistance of the thermistor. The lowered resistance allows current to increase, and the alarm is activated.

NTC Thermistor for Surge Protection

The circuit of Figure 5-7 shows an NTC thermistor acting as a shock absorber for a light bulb (incandescent lamp). When the circuit is energized, a high percentage of the available voltage is dropped across the thermistor. As the thermistor heats, its resistance drops and more current is available to the light bulb. Using this circuitry, many hours of life are added to a light bulb.

NTC Thermistor for Flow Measurement

An NTC thermistor dissipates heat at a constant rate under different environmental conditions. It can be used to measure flow as shown in Figure 5-8. Figure 5-8 shows a circuit with both thermistors directly heated. One thermistor is in the path of flow and the other is shielded from flow.

If the flow rate is large, the thermistor in path of the flow is much cooler than the other thermistor. Therefore, the output signal will be large. If the flow

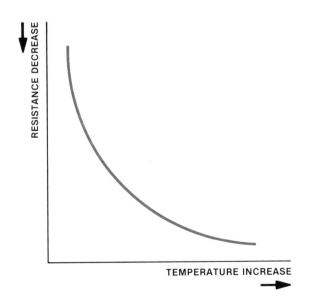

Figure 5-5. The resistance temperature characteristic curve for an NTC thermistor is approximately exponential (nonlinear). It has a large enough swing of resistance to become a versatile solid state device.

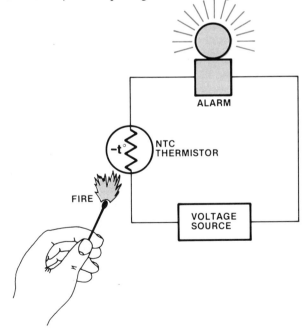

Figure 5-6. In the presence of fire, the increase temperature lowers the resistance of the NTC thermistor. The lowered resistance increases current and activates the alarm.

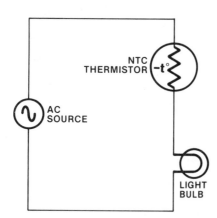

Figure 5-7. The NTC thermistor acts as a shock absorber for a light bulb by dropping a high percentage of the available voltage across the thermistor. As the thermistor heats, its resistance drops and more current is available to the light bulb.

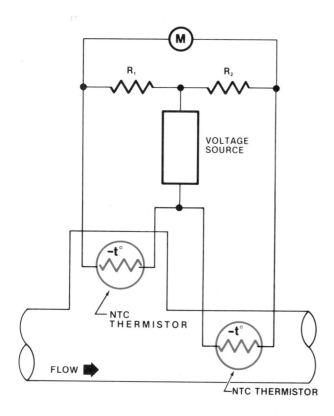

Figure 5-8. The amount of flow in this circuit determines how rapidly heat is removed from the NTC thermistor. Rapid flow dissipates heat rapidly. Less flow dissipates heat slowly. With a larger flow, there is less resistance and the signal is larger.

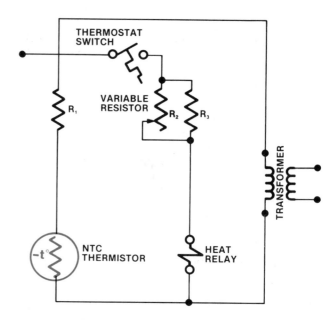

Figure 5-9. NTC thermistors used for temperature compensation in a thermostat can reduce the thermostat's variation in control temperatures to very low levels.

is not as rapid, heat is not carried away from the thermistor in the flow path as rapidly. The output signal would therefore be smaller.

NTC Thermistor for Temperature Compensation

NTC thermistors are commonly used for *temperature compensation*. When sensing outside ambient air temperatures, the thermistor is wired in series with resistor R_1 located inside a thermostat. See Figure 5-9. With a decrease in outdoor temperature, the NTC thermistor increases its resistance, allowing current to flow in the circuit. This reduces thermostat variations to improve the operating characteristics of a typical heat or heat pump system. Resistors R_2 and R_3 act as an additional heat anticipator whose resistance varies with the outdoor temperature.

PTC THERMISTOR APPLICATIONS

PTC thermistors are characterized by an extremely large resistance change for a small temperature span. See Figure 5-10. The temperature at which the resistance begins to increase rapidly is the *switch temperature*. This point can be changed from below zero to above 160 °C.

The PTC thermistor is similar to a switch that is not perfect, but is practical for many applications. A

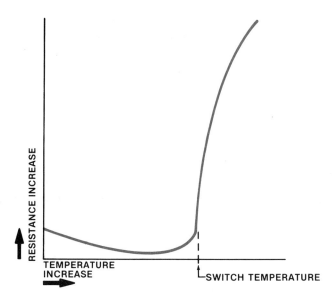

Figure 5-10. PTC thermistors have an extremely large resistance change for a small temperature span. The temperature at which the resistance begins to increase rapidly is the switch temperature.

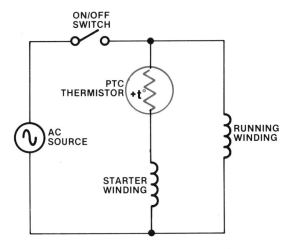

Figure 5-11. A PTC thermistor can be used to replace the troublesome start switch in a single-phase motor. As the resistance of the PTC thermistor increases, it causes a low current in the starter winding to effectively remove the starter winding from the circuit.

thermistor is not a perfect switch because it has some resistance when it is closed and a large amount of resistance when it is open. An ideal, or perfect switch, has no resistance when it is closed, and infinite resistance when it is open.

The PTC thermistor, depending upon the application, usually exhibits relatively low resistance levels during the ON state and relatively high resistance levels during the OFF state. Although current is always

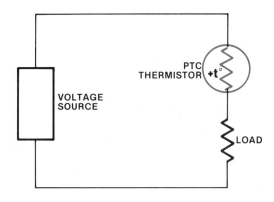

Figure 5-12. When a large current is presented to the load, the resistance of the PTC thermistor increases at a rapid rate so that a reduction in current occurs.

flowing in the circuit during the OFF state, it is usually so low that it is negligible.

PTC Thermistor for Motor Starting

In the circuit shown in Figure 5-11, the PTC thermistor replaces the start switch in a single-phase motor. When the circuit is energized by turning the ON/OFF switch ON, the PTC thermistor has a low resistance and permits most of the line voltage (AC source) to be applied to the starter winding. As the motor starts, the PTC thermistor heats up until the switch temperature is reached. This is determined by the thermal inertia of the PTC thermistor and current flowing through the starter winding. Then, the thermistor rapidly changes from a low-resistance device to a high-resistance device. Once the thermistor reaches its high-resistance state, virtually no current flows through the thermistor or the starter winding. Since the current through the starter is negligible, the starter winding can be considered removed from the circuit.

NOTE: The PTC thermistor becomes less effective on successive starts because its temperature is higher than the ambient temperature levels.

PTC Thermistor as a Current Limiter

The increase in resistance of a PTC thermistor at the switch temperature makes it suitable for current-limiting applications. See Figure 5-12. For currents lower than the limiting current, the power generated in the unit is insufficient to heat the PTC thermistor to its switch temperature. However, as the current increases to the critical level, the resistance of the PTC thermistor increases at a rapid rate so that any further increase in power dissipation results in a current reduction. The time required for the PTC thermistor to get into the current-limiting mode is controlled by the heat capacity of the PTC thermistor, its dissipation constant, and the ambient temperature.

PTC Thermistor for Arc Suppression

In the circuit of Figure 5-13, the PTC thermistor switches from a low resistance to a high resistance when the switch is open. The low resistance of the PTC provides effective arc suppression. In addition, the PTC switching action transfers essentially all of the power supply voltage from the load to the PTC thermistor itself.

TESTING THERMISTORS

The proper connection of a thermistor to any electronic circuit is very important. Loose or corroded connections create a high resistance in series with the thermistor resistance. The control circuit may sense the additional resistance as a false temperature reading.

The hot and cold resistance of a thermistor can be checked with an ohmmeter, but one end of the thermistor must be disconnected from the circuit. See Figure 5-14. The hot and cold resistance of a thermistor can then be tested as follows:

1. Connect the ohmmeter leads to the thermistor leads and place the thermistor and a thermometer in a mixture of ice and water.

2. Record the temperature and resistance readings.

3. Place the thermistor and thermometer in hot water (not boiling).

4. Record the temperature and resistance readings.

5. Compare these hot and cold readings with the manufacturer's specification sheet or a similar thermistor that is known to be good. When water may be harmful to the PC board or surrounding

components, a temperature testing unit should be substituted as the standard temperature source.

SOLID STATE PRESSURE SENSOR

A solid state *pressure sensor* is a transducer that changes resistance with a corresponding change in pressure. See Figure 5-15. The pressure sensor is designed to activate or deactivate when its resistance reaches a predetermined value. The pressure sensor is used for high or low pressure control depending on the switching circuit design. It is suited for a wide variety of pressure measurements on compressors, pumps, and other similar equipment. See Figure 5-16.

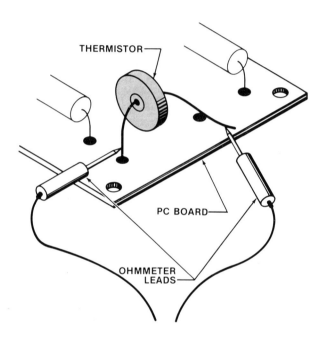

Figure 5-14. To check a thermistor, one lead must be disconnected from the circuit.

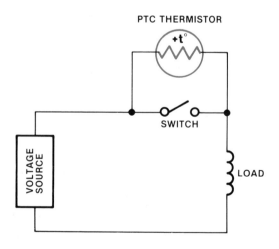

Figure 5-13. The PTC thermistor provides arc suppression. It absorbs the voltage from the inductive load by rapidly switching from a low resistance to a high resistance once the switch is open.

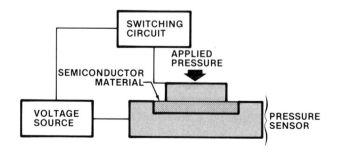

Figure 5-15. A solid state pressure sensor is a transducer that changes resistance with a corresponding change in pressure.

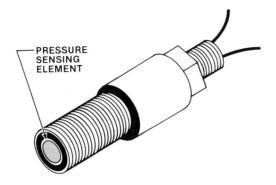

Figure 5-16. A pressure sensor can be used to measure pressure on compressors, pumps, and other similar equipment.

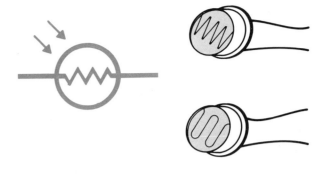

Figure 5-17. Although the schematic symbol for a photoconductive cell (left) is quite similar to that of a thermistor, the exposed nature of the semiconductor material (right) makes a photoconductive cell unmistakable by appearance.

A pressure sensor can detect low pressure, high pressure, or it can trigger a relief valve. Because a pressure sensor is extremely rugged, it is also used to measure compression in various types of engines. In some applications, the output signal of the transducer is applied to an oscilloscope so that the pattern on the oscilloscope provides a visual indication of changes in pressure. This is very valuable in analyzing engine performance.

Testing Solid State Pressure Sensors

The condition of a solid state pressure sensor can be tested as follows:

1. Disconnect the pressure sensor from the circuitry.

2. Connect the ohmmeter leads to the pressure sensor.

3. Activate the device being monitored (compressor, air tank, etc.) until pressure builds up. Record the resistance of the pressure sensor at the high pressure setting.

4. Open the relief or exhaust valve and reduce the pressure on the sensor. Record the resistance of the pressure sensor at the low pressure setting.

5. Compare these high and low resistance readings with manufacturer specification sheets. When specification sheets are not available, use a replacement pressure sensor that is known to be good.

6. As an additional check, the voltage supplied to the PC board and to the switching device should be checked while the circuit is in operation.

SEMICONDUCTOR PHOTOELECTRIC TRANSDUCERS

Depending upon the application, the major purpose of photoelectric devices is to produce a change of either resistance or voltage when exposed to light. Typically, photoelectric transducers are used as ON-OFF devices, measuring devices, or limited power sources.

An ON-OFF photoelectric transducer starts and stops a device because of changes in the light source. Examples of photoelectric transducers as ON-OFF devices are door openers, counters, and alarm systems. As a measuring device, the photoelectric transducer responds to changes in the intensity of light, or color of light. This results in a variable output voltage. Examples of photoelectric transducers as measuring devices are detectors, gas analyzers, and spectrometers. As a power source, photovoltaic cells are connected in series and parallel to provide voltage and current to support other electronic equipment. Examples of photoelectric transducers as power sources are remote radio transmitters and space satellites.

The two basic types of semiconductor photoelectric transducers are the *photoconductive cell* (photoresistive cell) and the *photovoltaic cell* (solar cell).

Photoconductive Cell

The schematic symbol for a photoconductive cell is similar to that of a thermistor. See Figure 5-17, left. The difference is the absence of the letter *t* and the addition of two arrows. The construction of a typical photoconductive cell is shown in Figure 5-17, right. The photoconductive cell is formed with a thin layer of semiconductor mateial such as cadmium sulfide or cadmium selenide deposited on a suitable insulator. Leads are attached to the semiconductor material and the entire assembly is hermetically sealed with glass. The transparency of the glass allows light to reach the semiconductor material.

For maximum current-carrying capacity, the photoconductive cell is manufactured with a short conduction path having a large cross-sectional area.

If a number of photoconductive cells are connected in parallel, they can produce up to one-half amp.

Photoconductive Cell Operation. The resistance of a photoconductive cell changes as the light striking its surface changes. When light strikes a photoconductive cell, electrons are freed, and the resistance of the material is lowered. When the light is removed, the electrons and holes recombine. A photoconductive cell ranges from several megaohms in total darkness to less than one hundred ohms at full light intensity.

NOTE: A certain amount of current (dark current) will flow even when it is dark. Dark current has no effect on a photoconductive cell used for ON-OFF operations. It may, however, have to be taken into account in sensitive measuring circuits where small amounts of current would distort the readings.

Photoconductive Cell Applications. Photoconductive cells are used when time response is not critical. They would not be used where several thousand responses per second are needed to transmit accurate data. However, they can be used efficiently with slower responding electromechanical equipment such as pilot lights or street lights.

Pilot Light. Figure 5-18 shows a circuit in which a photoconductive cell is appropriate. In this case, the light level determines if the pilot light (flame) on a gas furnace is ON or OFF. When a pilot light is present, the light from the flame reduces the resistance of the photoconductive cells. Current is allowed to pass through the cell and activate a control relay. The control relay in turn allows the main gas valve to be energized when the thermostat calls for heat.

Street Light. Figure 5-19 shows a street light circuit in which a photoconductive cell is used. An increase of light at the photoconductive cell results in a decrease in resistance, and current is increased through the relay. The increased current in the relay causes the normally closed (NC) contacts to open, and the light turns OFF. With darkness, the resistance increases, causing the NC contacts to return to their original position, and the light goes ON.

Testing Photoconductive Cells. Humidity and contamination are the primary causes of photoconductive cell failure. The use of quality components that are hermetically sealed is essential for long life and proper operation. Some plastic units are less rugged and more susceptible to temperature changes than glass units.

The resistance of a photoconductive cell is tested with an ohmmeter, but the photoconductive cell must have one lead disconnected from the circuitry:

1. Connect the ohmmeter leads to the photoconductive cell.

2. Cover the photoconductive cell and record its dark resistance.

3. Shine a light on the photoconductive cell and record its light resistance.

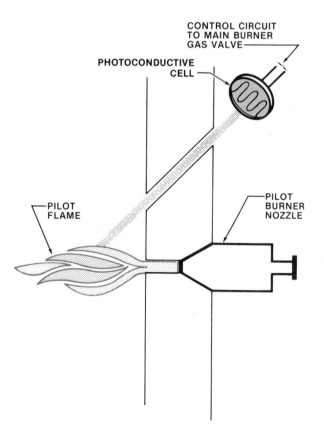

Figure 5-18. The photoconductive cell is used to determine if the pilot light on the gas furnace is ON or OFF.

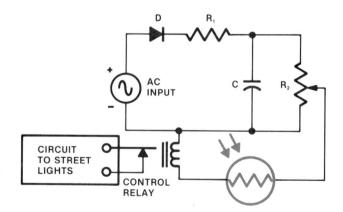

Figure 5-19. The photoconductive cell is used to determine when the street light should turn ON or OFF. Diode *D*, resistor R_1, and capacitor *C* form a half-wave rectifier that supplies power to the coil of the control relay. Resistor R_2 determines the output of the supply to the coil.

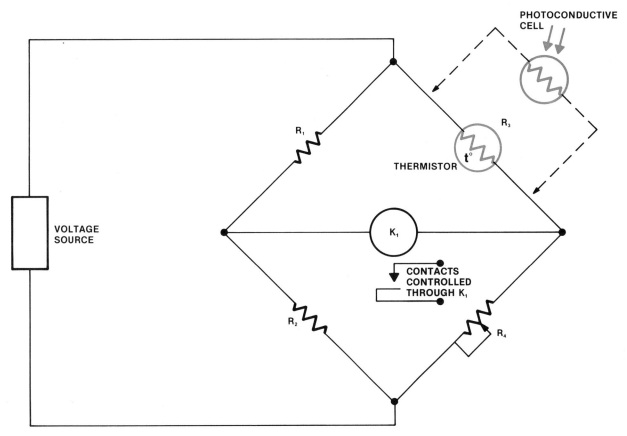

Figure 5-20. With a bridge circuit, more precise measurements and control can be obtained compared to series or parallel connections.

4. Compare these resistance readings with the manufacturer's specification sheets. When specification sheets are not available, use a similar photoconductive cell that is known to be good.

As with other resistance-type sensors, the connection to the control circuit is very important. All connections should be tight and corrosion-free.

Bridge Circuits with Sensing Devices. Resistance can be arranged to form a bridge circuit. See Figure 5-20. With a bridge circuit, more precise measurements and control can be obtained compared to a series or parallel connection. A sensing device (photoconductive cell or thermistor) is placed in one arm of the bridge so that it can detect changes in light (for a photoconductive cell) or temperature (for a thermistor). Resistance R_3 in this case represents the sensing device. The bridge current is then balanced through the use of the potentiometer R_4. Resistors R_1 and R_2 are fixed-value precision resistors.

When the bridge is balanced, there will be no current across the bridge circuit. Because there is no current, the relay coil K_1 in this circuit is de-energized. Any changes in light or temperature will change the resistance of R_3 and the bridge becomes unbalanced. With the bridge unbalanced, current flows through K_1 and the relay becomes energized.

If a galvanometer is substituted for the relay coil of the bridge circuit, the circuit becomes an extremely sensitive light meter or thermometer. See Figure 5-21. If a photoconductive cell is used, it can then measure very minute changes in light. If a thermistor is used, it can measure very minute changes in temperature.

Many industrial bridge circuits contain two solid state thermistors. See Figure 5-22. Here, two matched thermistors are used to form a measuring circuit. Normally, one thermistor is placed where its temperature is kept at a fixed value, and the other thermistor is used as a monitor. As long as the two thermistors have equal temperatures, the bridge circuit is balanced and no current flows across the bridge. However, should the monitoring thermistor detect an increase or decrease in temperature, then current would flow across the bridge. This current could then be used to activate a relay, drive a meter movement, or serve as a control circuit within another circuit.

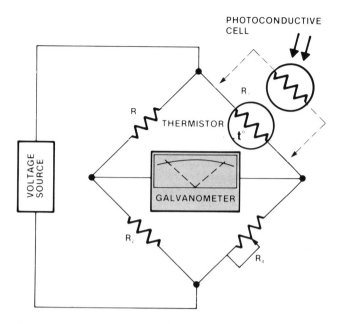

Figure 5-21. When a very sensitive meter movement (galvanometer) is placed across a bridge circuit, minute changes can be measured.

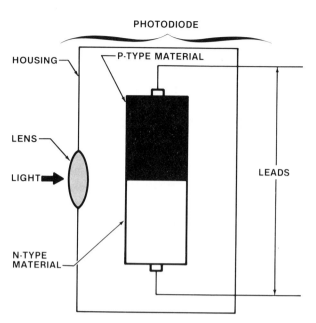

Figure 5-23. The primary difference between a regular semiconductor diode and a photodiode is the addition of a lens for focusing light on the PN junction area.

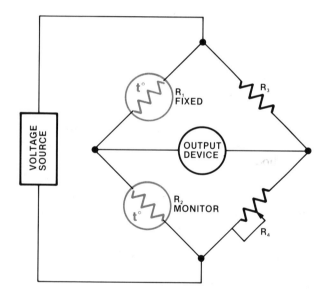

Figure 5-22. When two thermistors are used in a bridge circuit, they are a matched set. One is kept at a fixed value and the other is used as a monitor.

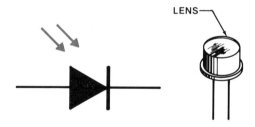

Figure 5-24. The photoconductive diode symbol differs from the regular semiconductor diode in that two small arrows are added. The photoconductive diode also has a lens in its housing.

Photoconductive Diode

A *photoconductive diode (photodiode)* is internally much the same as a regular semiconductor diode. The primary difference is the addition of a lens in the housing for focusing light on the PN junction area. See Figure 5-23. The schematic symbol for a photoconductive diode is shown in Figure 5-24 along with a housing for photodiodes.

Photoconductive Diode Operation. With the photodiode, the conductive properties change when light strikes the surface of the PN junction. Without light, the resistance of the photodiode is high. When it is exposed to light, the resistance reduces proportionately.

NOTE: The photodiode is usually operated in the reverse bias mode. See Figure 5-25.

Photodiode Applications. Photodiodes respond much faster than photoconductive cells, and they are usually more rugged. Photodiodes are found in movie equipment, punched tape readers, and other equipment requiring a rapid response time.

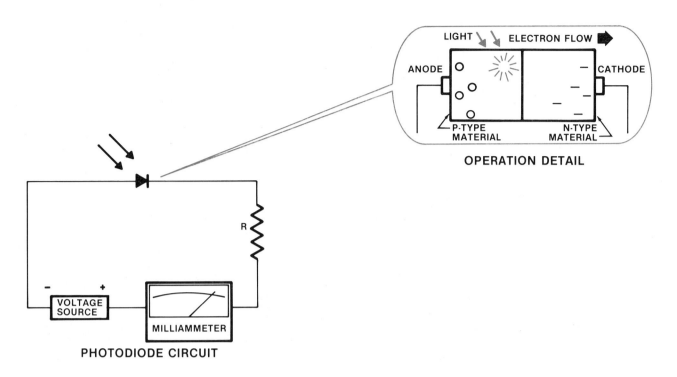

OPERATION DETAIL

Figure 5-25. The photodiode is normally operated in the reverse bias mode.

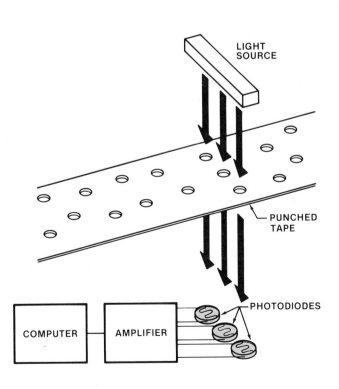

Figure 5-26. Photodiodes are used to read the information contained in a punched tape.

Optical Readout Device. Photodiodes can be used for reading information on computer numerical control (CNC) machine tapes. Figure 5-26 illustrates a simple optical readout device for a CNC machine.

A constant light source is placed above the photodiodes so that a punched tape can move between the light source and the photodiodes. The photodiodes are not energized as long as there are no holes in the tape to allow light to pass. When a hole passes under the light, the corresponding light energizes the photodiode and the response is amplified and recorded in the computer network. This arrangement eliminates the need for mechanical readout equipment, which is much slower.

Sensing Devices. Photodiodes are also used for a variety of sensing applications. See Figure 5-27. They are used to count, measure diameters, maintain level, and read coded marks for initiating operations in an automated process. Photodiodes can also be used to optically isolate circuits from noise and interference.

Photovoltaic Cell

The function of a photovoltaic cell (solar cell) is to convert solar energy to electrical energy. A photovoltaic cell is sensitive to light and it produces a voltage without an external source. There are several different types of photovoltaic cells. See Figure 5-28.

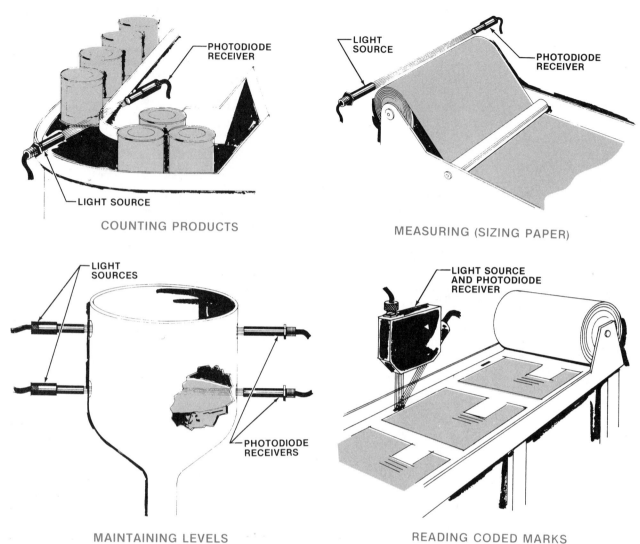

Figure 5-27. Photodiodes are used to count, measure, maintain level, and read coded marks. *(Micro Switch, A Division of Honeywell.)*

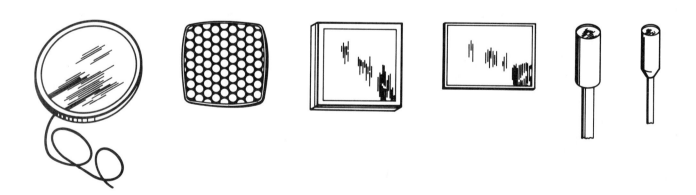

Figure 5-28. The photovoltaic cell (solar cell) is sensitive to light and produces a voltage without an external source.

The schematic symbol for the photovoltaic cell is shown in Figure 5-29. The symbol indicates that the device is equivalent to a single-cell voltage source like those found in batteries.

The use of photovoltaic cells is increasing. As they are perfected and decrease in cost, their use as a remote power source will become more popular. Many manufacturers are designing into their products, the use of photovoltaic cells on individual and multi-cell applications.

Photovoltaic Cell Operation. The photovoltaic cell generates energy by using a PN junction to convert light energy into electrical energy. See Figure 5-30. It produces a potential difference between a pair of terminals only when exposed to light.

At the junction of N-type material and P-type material, some recombination of electrons and holes occurs, but the junction itself acts as a barrier between the two types of material. The electrical field at the

junction maintains the negative charges on the N-type material side and the holes or positive charges on the P-type material side.

If the load is connected across the PN junction, current flows with the light acting as a generator. When current flows through the load, electron-hole pairs formed by light energy recombine and return to the normal condition prior to the application of light. Consequently, there is no loss or addition of electrons to the silicon during the process of converting light energy to electrical energy. The photovoltaic cell should have no limit to its life span, provided it is not damaged.

Photovoltaic Cell Output. Photovoltaic cells are rated by the amount of energy they convert. Most manufacturers rate the output in terms of volts (V) and milliamps (mA). Photovoltaic cells may operate with up to 0.5 volts and as high as 40 milliamps (.040 amps) per cell. To increase the current output from a set of photovoltaic cells, they should be placed in parallel. See Figure 5-31. To increase the voltage output, they should be placed in series. See Figure 5-32.

Photovoltaic Cells for Light-Intensity Meters. One of the best known applications for the photovoltaic cell is as a light-intensity meter, or exposure meter, used in photography. See Figure 5-33. The light reflected from the subject to be photographed passes through the rectangular window and strikes the photovoltaic cell. A sensitive meter movement is connected across the output of the photovoltaic cell and the current flows in proportion to the amount of light falling on it. The meter reading is used in conjunction with information regarding the various types of film available. This determines shutter speed and lens opening used for taking the picture.

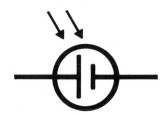

PHOTOVOLTAIC CELL (SOLAR CELL)

Figure 5-29. The schematic symbol for a photovoltaic cell is equivalent to a single-cell voltage source like those found in batteries.

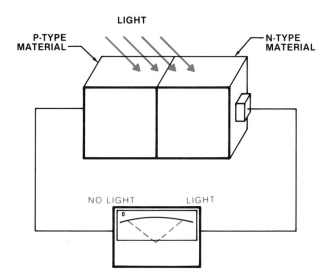

Figure 5-30. The photovoltaic cell will produce a potential difference (voltage) between a pair of terminals only when exposed to light.

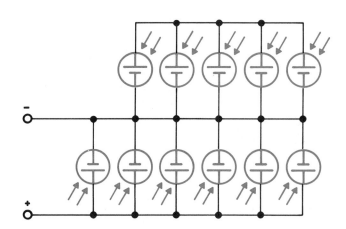

Figure 5-31. To increase the current output from a set of photovoltaic cells, they should be placed in parallel.

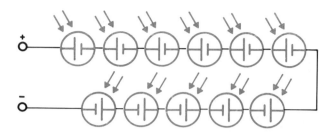

Figure 5-32. To increase the voltage output from a set of photovoltaic cells, they should be placed in series.

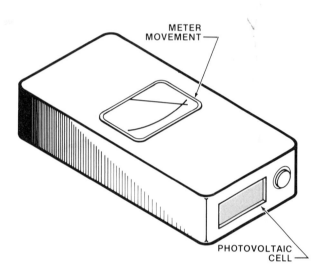

Figure 5-33. The light meter uses the photovoltaic cell to determine the intensity of the light.

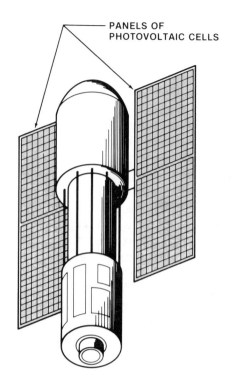

Figure 5-34. Weather satellites use photovoltaic cells to produce electrical energy.

Photovoltaic Cells for Satellite Power Sources. Figure 5-34 shows a weather satellite using two panels of photovoltaic cells for producing electrical energy. The electrical systems of satellites are powered by solar energy. The solar cells convert light rays of the sun into electrical energy. Communication satellites, which operate almost indefinitely in space, have hundreds of such solar cells constantly converting solar energy into electrical energy. Satellites use the electrical energy to receive and transmit television and radio signals to and from the surface of the earth.

HALL EFFECT SENSORS

The *Hall effect sensor* was discovered by Edward H. Hall in 1879 at John Hopkins University. It was not until the steady growth of semiconductor technology and their mass production that Hall effect sensors were used in factories as well as in laboratories. Since then, Hall effect sensors have been used in many products such as computers, sewing machines, automobiles, aircraft, machine tools, and medical equipment.

Hall Effect Principle

The Hall effect principle is shown in Figure 5-35. A constant control current passes through a thin strip of semiconductor material (*Hall generator*). When a permanent magnet is brought near, a small voltage, called Hall voltage, appears at the contacts that are placed across the narrow dimension of the strip. As the magnet is removed, the Hall voltage reduces to zero. Thus, the Hall voltage is dependent on the presence of a magnetic field and on the current flowing through the Hall generator. If either the current or the magnetic field is removed, the output of the Hall generator would be zero. In most Hall effect sensors, the control current is held constant and the flux density is changed by movement of a permanent magnet.

NOTE: The Hall generator must be combined with an association of electronic circuits to form a Hall effect sensor. Since all of this circuitry is usually on an integrated circuit (IC), the Hall effect sensor can be considered a single device with a voltage output.

Hall Effect Sensor Packaging

Hall effect sensors are packaged in different ways and are chosen according to the type of job they are to

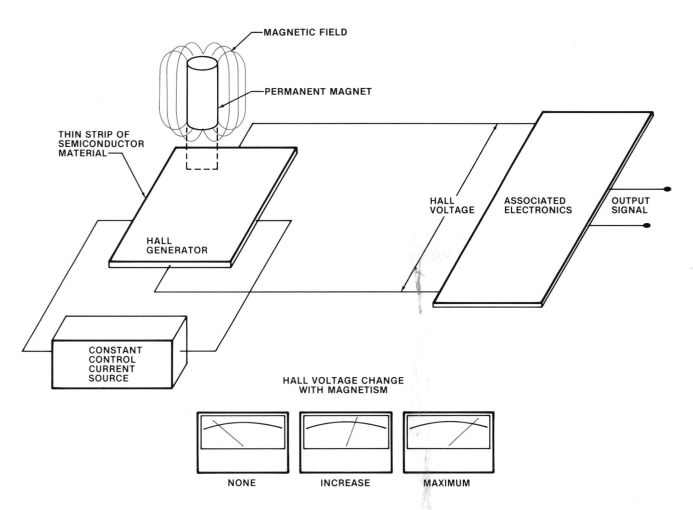

Figure 5-35. The output of the Hall generator depends upon the presence of a magnetic field and the current flow in the Hall generator.

perform. Figure 5-36 shows five of the most common packaging arrangements: vane, plunger, current sensor, proximity, and cylindrical.

Hall Effect Sensor Applications

Hall effect sensors are used in a variety of commercial and industrial applications. Their size, light weight, and ruggedness make them ideal for many sensing jobs that are impossible to accomplish with other types of devices. For example, Hall effect sensors can be embedded in the human heart to serve as the timing element.

Most of the following Hall effect sensor functions can be accomplished by using either a digital or linear output transducer. The choice depends on output requirements for each application:

Under or overspeed detection
Disk speed detection
Automobile or tractor transmission controller

Shaft rotation counter
Bottle counting
Camera shutter positioning
Rotary position sensing
Flow-rate meter
Tachometer pick-ups

RPM Sensors. The RPM sensor is one of the most common applications of a Hall effect sensor. The magnetic flux required to operate the sensor may be furnished by individual magnets mounted on the shaft or hub, or by a ring magnet. Each change in polarity results in an output signal. Figure 5-37 illustrates some basic concepts for designing the RPM sensor.

Door Sensor. Figure 5-38 illustrates the use of a Hall effect sensor as a door-interlock sensing device for an electronic enclosure. The sensor is recessed in the door frame and a magnet is embedded in the door. When the door is closed, the magnet actuates the sensor, and current flows through the door-sensing circuit. When

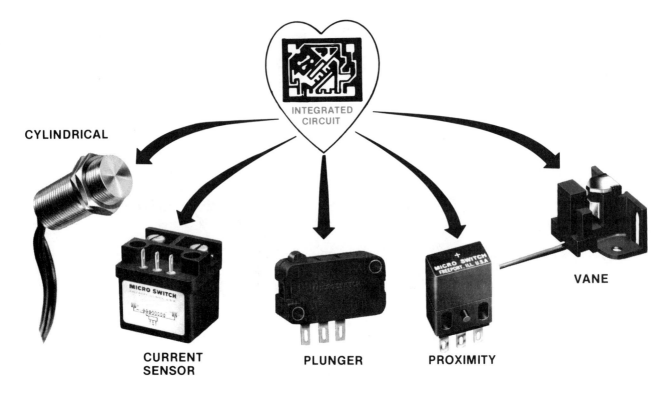

Figure 5-36. Hall effect sensors are packaged in several ways. *(Micro Switch, A Division of Honeywell.)*

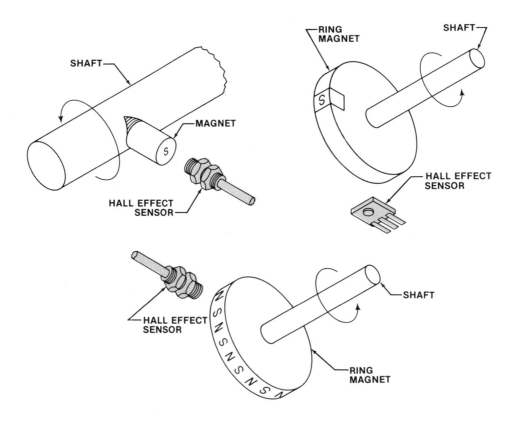

Figure 5-37. Each change in polarity results in an output from the Hall effect sensor used in an RPM sensor application. *(Micro Switch, A Division of Honeywell.)*

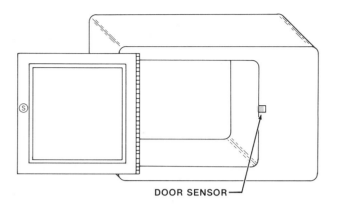

Figure 5-38. When the door is closed, the magnet actuates the sensor, and current flows through the door-sensing circuit. When the door opens, the magnet moves away from the sensor, and it is returned to its normally OFF condition. *(Micro Switch, A Division of Honeywell.)*

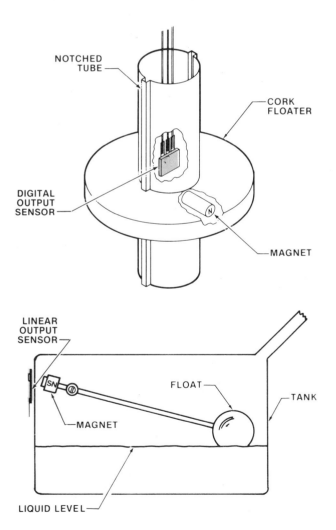

Figure 5-39. The Hall effect sensor can be used to measure liquids in a tank. *(Micro Switch, A Division of Honeywell.)*

the door opens, the magnet moves away from the sensor and it is returned to its normally OFF condition. This system requires connecting wires only in the door frame, not in the door.

Low-Fuel Warning Sensor. Figure 5-39 shows two methods for measuring the level of liquids in a tank. One method uses a notched tube with a cork floater inserted into the gas tank. See Figure 5-39, top. The magnet is mounted in the float assembly, which is forced to move in one plane (up and down). As the liquid level goes down, the magnet passes the digital output sensor. When the sensor is actuated, this could indicate a low fuel condition. A linear output sensor may be used to indicate fuel level.

Figure 5-39, bottom, illustrates another method that uses a linear output sensor and a float in a tank made of aluminum. As the liquid level goes down, the magnet moves closer to the sensor, causing an increase in output voltage. This method allows measurement of liquid levels without any electrical connections inside the tank.

Magnetic Card Reader. A door-interlock security system can be designed using a Hall effect sensor, a magnetic card, and associated electronic circuitry. See Figure 5-40. The magnetic card slides by the sensor and produces an output signal. This analog signal is converted to digital to provide a crisp signal to energize the door latch relay. When the solenoid of the relay pulls in, the door is opened. For systems that require additional security measures, a series of magnets may be molded into the card.

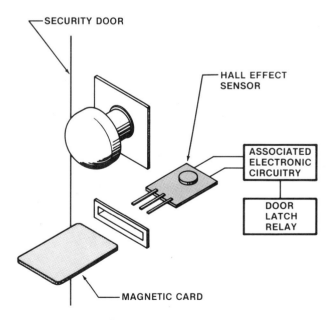

Figure 5-40. A door-interlock system can be designed using a Hall effect sensor, a magnetic card, and associated electronic circuitry.

LIGHT EMITTING DIODES

The *light emitting diode (LED)* produces light through the use of semiconductor materials. A diode junction can emit light when an electrical current is present. The presence of an electrical current produces light energy because electrons and holes are forced to recombine.

The energy level of an electron as it passes through the junction of a semiconductor diode can be shown graphically. See Figure 5-41. To get through the depletion region, the electron must aquire some additional energy. This additional energy comes from the positive field of the anode. If the field is not strong enough, the electron will not get through the depletion region and no light will be emitted. For a standard silicon diode, a minimum of 0.6 volts must be present before the diode will begin to conduct. For a germanium diode, 0.3 volts must be present before the diode will begin to conduct. Most LED manufacturers make a larger depletion region that requires 1.5 volts for the electron to get across the depletion region. As the electron moves across the depletion region, it gives up its extra kinetic energy. The extra energy is converted to light.

LED Construction

Manufacturers of LEDs generally use a combination of gallium and arsenic with silicon or germanium to construct semiconductors. By adding and adjusting other impurities to the base semiconductor, different wavelengths of light can be produced. LEDs are capable of producing a light that is not visible to the human eye, called *infrared light*, or it may emit a visible red or green light. If other colors are desirable, the plastic lens may be of a different color.

Like standard semiconductor diodes, there must be a method for determining which end of an LED is the anode and which end is the cathode. The cathode lead is identified by the flat side of the device, or it may have a notch cut into the ridge. See Figure 5-42. The schematic symbol for an LED is exactly like that of a photodiode, but the arrows point away from the diode. See Figure 5-43. The LED must be forward biased, and a current-limiting resistor is usually present to protect the LED from excessive current.

The colored plastic lens magnifies the light produced at the wire of the LED. See Figure 5-44. Without the lens, the small amount of light produced at the wire would be diffused, and it would be virtually unusable as a light source. The size and shape of the LED package determines how it must be positioned for proper viewing.

LED Applications

As a limited light source, the LED can be used as a substitute for applications once reserved for small incandescent bulbs. Since the LED is inexpensive, has an unlimited life span, and is a low power comsumption device, it is ideal for a variety of applications.

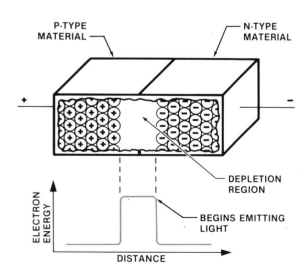

Figure 5-41. As the electron moves across the depletion region, the electron gives up its extra kinetic energy. The extra energy is converted to light.

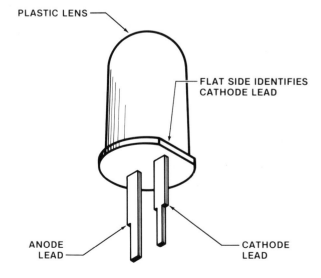

Figure 5-42. When color is required from an LED, a plastic lens may be added. The flat side (or notch) identifies the cathode lead.

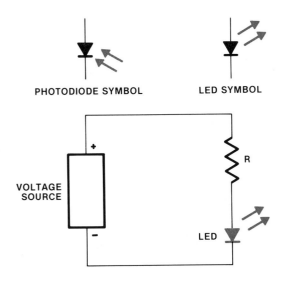

Figure 5-43. The schematic symbol for an LED is exactly like a photodiode except for the direction of the arrow. The LED must be forward biased, and a current-limiting resistor *R* is usually present to protect the LED from excessive current.

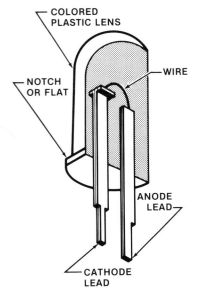

Figure 5-44. The lens of an LED magnifies the light produced at the wire.

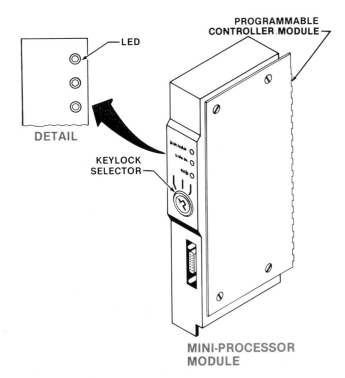

Figure 5-45. When used as a status indicator, the LED lets the operator of a machine know something about the operating status of the machine.

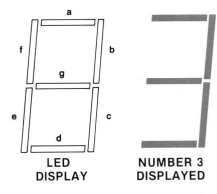

Figure 5-46. When LEDs are combined, they can form a seven-segment display. By energizing the proper sequence, a number can be created. For example, number 3 is formed by energizing segments a, b, g, c, and d.

Status Indicator. A *status indicator* is an LED application that lets the operator of a machine know something about the status of the machine's operation. It may be used to indicate whether it is ON or OFF, or whether it is malfunctioning. In the case of a mini-processor module with a programmable controller, the LEDs can be used for diagnostics. See Figure 5-45. In this case, the LEDs will indicate whether the machine is running properly (RUN), or whether there is a problem in the processor (PROCESSOR), or in the memory (MEMORY). Selecting the diagnostic output is accomplished by using a keylock selector.

Seven-Segment Display. The readout on test equipment is often presented on a seven-segment display. The display is made with LEDs and is called an LED display. Each segment of the display has several LEDs, but a colored plastic overlay makes the individual LEDs glow as a unit. The number 3 would be activated on the display by energizing segments a, b, g, c, and d. See Figure 5-46.

5 × 7 Dot-Matrix Display. It is often necessary to show symbols and letters of the alphabet as well as numbers. To accomplish this, 5 × 7 dot-matrix displays are used. See Figure 5-47, left. This type of display uses one LED for each dot of the matrix. For example, when the proper sequence is energized, the letter E will appear. This type of display is often called an *alphanumeric display* since it can create letters, numbers, and symbols.

LIQUID CRYSTAL DISPLAYS (LCD)

A *liquid crystal display (LCD)* is packaged much like an LED and it can produce the same numeric and alphanumeric information. Two significant differences between the devices are that the LCD controls light while the LED generates light, and the LCD requires much less power to operate than an LED. Because of these advantages of the LCD, it is used extensively in newer test equipment, clock displays, and calculators.

LCD Construction

An LCD consists of a front and rear piece of glass. These two pieces of glass are separated by a *nematic liquid* (liquid crystal material). See Figure 5-48. Both pieces of glass are coated with a microscopically thin layer of metal that is transparent. The coating is applied to each piece of glass so that it faces the nematic liquid. The layer of metal applied to the front surface of the rear piece of glass covers the entire active area of the display. The layer of metal applied to the rear surface of the front piece of glass is broken into segments. The metal segments are then brought through the separator seal to the edges of the display to provide electrical connection points for the driving circuitry.

The key to LCD operation is the nematic liquid that fills the space between the front glass and the rear

glass. The molecules of the nematic liquid are normally parallel to the plane of the glass. However, when a voltage is applied to the nematic liquid, its molecules twist 90° to alter the light passing through it. See Figure 5-49.

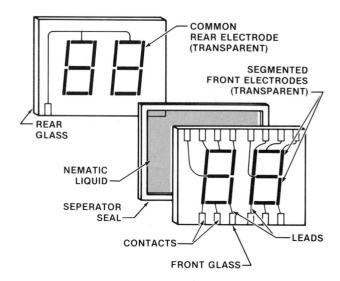

Figure 5-48. A liquid crystal display (LCD) is composed of two pieces of glass separated by a nematic liquid.

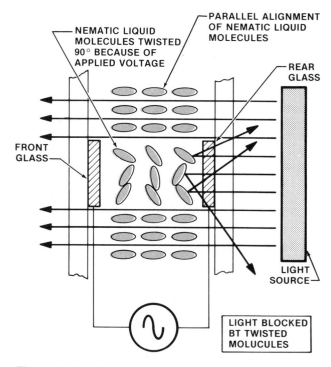

Figure 5-49. The molecules of the nematic liquid are normally parallel to the plane of the glass. However, when a voltage is applied to the nematic liquid, its molecules twist 90° to alter the light passing through it.

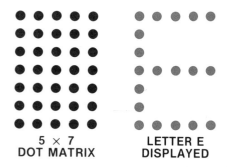

5 × 7
DOT MATRIX

LETTER E
DISPLAYED

Figure 5-47. When one LED is used to represent a dot in a display, a dot-matrix is formed.

By using *polarizing filters* (filters that allow light to pass only at certain angles), it is possible to obtain black digits or clear digits with an LCD. The only design change between the two types of displays is a 90° rotation of one of the polarizing filters. Users of LCDs generally select the type of digits best suited for their equipment application. Wristwatch manufacturers use displays with black digits because of the sharp contrast they have against the light-colored background. Clear digits are often preferred in digital-panel meters where black lighting (usually through a milky-white or colored filter) is practical and desirable.

Another advantage of the LCD over an LED is that it is less prone to wash-out under strong ambient light. The digits of an LCD actually become a deeper black, or they become clearer, as the intensity of the ambient light increases.

The disadvantage of LCDs, however, is that they cannot be read in the dark and must be illuminated by an external light source. This disadvantage is usually overcome by installing a small light source near the display. The user can switch the light ON and OFF as needed to read the display.

LCD Operating Voltages

All LCDs require special operating voltages. Applying improper voltages can drastically reduce the life of the display. Improper voltages can also make the display inoperative. The proper operating voltage for any LCD is an AC voltage with no DC present. Using a DC voltage, or an AC voltage with DC present, reduces the life of the display, often to only a few days. When the proper AC voltage is used, the life of the display may be several years. The AC voltage applied to the display need not be sinusoidal. In fact, it is often in the form of a square wave.

Chapter 5 - Review Questions

1. What is a transducer?
2. What is a thermistor?
3. Can thermistors be operated on AC and DC?
4. What is the difference between positive temperature coefficient (PTC) and negative temperature coefficient (NTC) in regard to thermistors?
5. Explain the difference between hot and cold resistance.
6. Describe the procedure for testing a thermistor.
7. Give two examples of PTC and NTC applications.
8. How does a pressure sensor work?
9. What three applications are photoelectric transducers usually used for?
10. What is the difference between a photoconductive cell and a photovoltaic cell?
11. What type of circuit can be used to make a transducer more sensitive?
12. What is a photodiode?
13. How does a photovoltaic cell operate?
14. Explain the Hall effect principle.
15. Name four applications for Hall effect sensors.
16. How do light emitting diodes (LEDs) operate?
17. How are different colored LEDs constructed?
18. What is a status indicator?
19. What is an alphanumeric display?
20. How is an LCD different from an LED?

6 TRANSISTOR AS DC SWITCH

The two types of transistors are PNP and NPN. Figure 6-1 shows their basic construction and their corresponding schematic symbol. The PNP transistor is formed by sandwiching a thin layer of N-type material between two layers of P-type material. The NPN transistor is formed by sandwiching a thin layer of P-type material between two layers of N-type material. Transistors are three-terminal devices. Their terminals are the *emitter (E), base (B),* and *collector (C)*. The emitter, base, and collector are located in the same place for both symbols. The only difference is the direction in which the emitter arrow points. However, in both cases, the arrow points from the P-type material toward the N-type material. (The term *bipolar* is often used when describing a transistor. It means that both holes and electrons are used as internal carriers for maintaining current flow in the transistor.)

Key Words

Base	Cutoff region	Negative feedback
Base-collector junction	Degenerative feedback	Saturation region
Base-emitter junction	Delta (Δ)	Seven-segment display readout
Beta (β)	Emitter	Thermal instability
Bias stabilization circuits	Emitter feedback	TO number
Bipolar	h_{fe}	V_{BB}
Collector	I_B	V_{CB}
Collector feedback	I_C	V_{CC}
Combination bias	Index pin	V_{CE}
Current gain	Load line	

TRANSISTOR TERMINAL ARRANGEMENTS

Transistors are manufactured with either two or three leads extending from their case. See Figure 6-2. These packages are accepted industry-wide no matter which company manufactures them. When a specific shaped transistor must be used, its *transistor outline (TO) number* is used as a reference. See Figure 6-3. TO numbers are determined by individual manufacturers.

NOTE: The bottom view of transistor TO-3 shows only two leads (terminals). Frequently, transistors use the metal case as the collector-pin lead.

Spacing can also be used to identify leads. Usually, the emitter and base leads are close together and the collector lead is further away. The base lead is always in the middle. See Figure 6-4. A transistor with an *index pin* must be viewed from the bottom. See Figure 6-5. The leads are identified in a clockwise direction from the index pin. The emitter is closest to the index pin.

NOTE: For detailed information on transistor construction and identification, refer to a transistor manual or to the manufacturer's specification sheets.

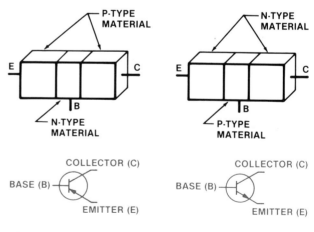

Figure 6-1. The two types of transistors are PNP and NPN.

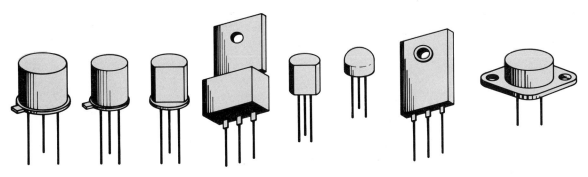

Figure 6-2. Transistors have either two or three leads extending from their case.

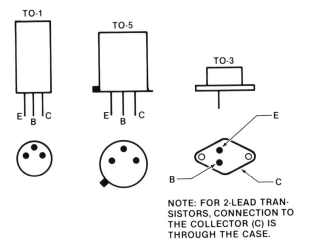

Figure 6-3. When a specific shaped transistor must be used, its TO number is used as a reference.

NOTE: FOR 2-LEAD TRANSISTORS, CONNECTION TO THE COLLECTOR (C) IS THROUGH THE CASE.

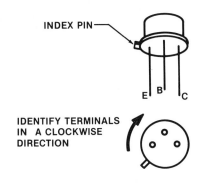

Figure 6-5. The transistor must be viewed from the bottom when an index pin is used to identify leads. The leads are identified in a clockwise direction from the index pin. The emitter lead is closest to the index pin.

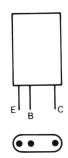

Figure 6-4. Spacing can be used to identify leads. The emitter and base leads are close together while the collector lead is further away. The base lead is always in the middle.

BIASING TRANSISTOR JUNCTIONS

In any transistor circuit, the *base-emitter junction* must always be forward biased and the *base-collector junction* must always be reverse biased. Figure 6-6 shows the base-emitter junction of an NPN transistor.

The external voltage (bias voltage) is connected so that the positive terminal connects to the P-type material (the base), and the negative terminal connects to the N-type material (the emitter). This arrangement forward biases the base-emitter junction. Current flows in the external circuit as indicated by the arrows. The action that takes place is the same as the action that occurs for the forward biased semiconductor diode.

Figure 6-7 illustrates the base-collector junction of an NPN transistor. The external voltage is connected so that the negative terminal connects to the P-type material (the base) and the positive terminal connects to the N-type material (the collector). This arrangement reverse biases the base-collector junction. Only a very small current (leakage current) will flow in the external circuit as indicated by the dashed arrows. The action that takes place is the same as the action that occurs for a semiconductor diode with reverse bias applied.

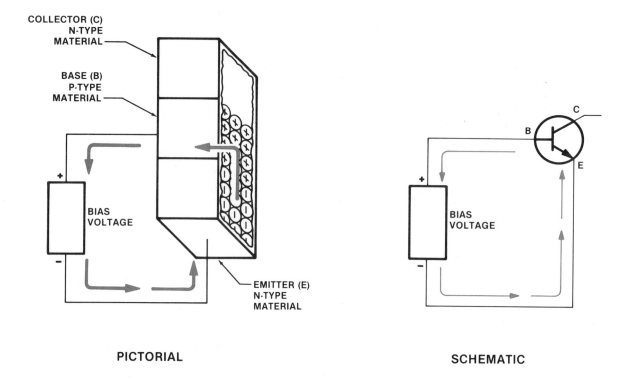

PICTORIAL   SCHEMATIC

Figure 6-6. When the base-emitter junction of this NPN transistor is forward biased, current flows in the direction indicated by the arrows in the external circuit.

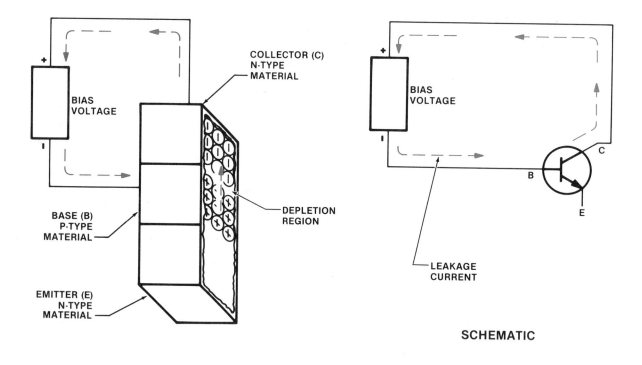

PICTORIAL   SCHEMATIC

Figure 6-7. When the base-collector circuit of this NPN transistor is reverse biased, only a very small current (leakage current) flows in the external circuit as indicated by the dashed arrows.

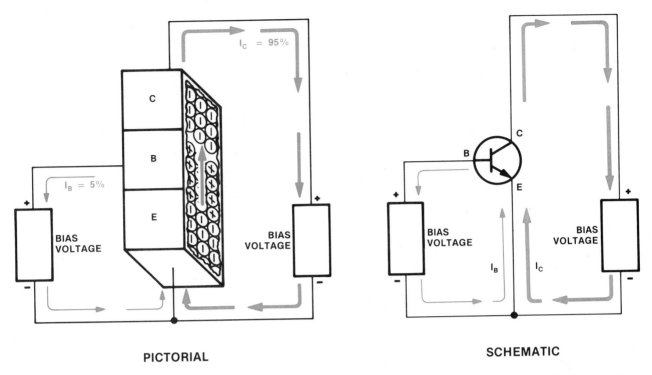

PICTORIAL   SCHEMATIC

Figure 6-8. With both junctions biased, an entirely different current path is created than with the junctions biased separately. Because the base is so thin, the more positive potential of the collector pulls the electrons through the thin base material. The result is that 95% of the current is in the collector and only 5% goes through the base.

Transistor Current Flow

Individual PN junctions can be used in combination with two bias arrangements. See Figure 6-8. The base-emitter junction is forward biased while the base-collector junction is reverse biased. This circuit arrangement results in an entirely different current path than the path that occurs with the individual circuits only. The heavy arrows indicate that the main current path is directly through the transistor. The main current path has a small amount of current through the base as indicated by the thinner arrows.

Figure 6-8 shows that the forward bias of the emitter-base circuit causes the emitter to inject electrons into the depletion region between the emitter and the base. Because the base B is thin (less than .001″ for most transistors), the much more positive potential of the collector C pulls the electrons through the thin base. As a result, the greater percentage (95%) of the available free electrons (I_C) from the emitter passes directly through the base into the N-type material, which is the collector of the transistor.

Control of Base Current

The base current I_B is a critical factor in determining the amount of current flow in a transistor. It is critical because the forward biased junction has a very low resistance and could be destroyed by heavy current flow. Therefore, the base current must be limited and controlled.

One method of limiting the base current in a transistor is with a series-limiting resistor R_1. See Figure 6-9. R_1 is extremely important because it keeps the base-emitter junction voltage and current from getting too high. The voltage applied to the

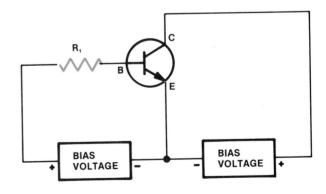

Figure 6-9. The series-limiting resistor R_1 forms a voltage divider with the resistance of the base-emitter junction to reduce the voltage and current in the base-emitter junction. Voltage for a silicon transistor is 0.6 volt and 0.3 volt for a germanium transistor.

base-emitter junction must be very small (0.6 volts for silicon and 0.3 volts for germanium). R_1 forms a voltage divider with the resistance of the base-emitter junction to accomplish this voltage reduction. Since R_1 is considerably larger in resistance compared to the base-emitter junction, only a small fraction of the voltage applied will be across the base-emitter junction. This type of voltage biasing is called fixed biasing since resistor R_1 does not change in value.

Voltage in a Transistor

Figure 6-10 shows the voltages associated with the transistor in the base circuit. V_{BB} indicates bias voltage applied to the base-emitter circuit. V_{BE} indicates the voltage drop across the base-emitter junction. V_{R1} indicates the voltage drop across the series-limiting resistor R_1. Since the resistance of the base-emitter junction is less than R_1, most of the applied voltage (V_{BB}) will be across R_1.

Figure 6-11 shows the other voltages associated with the transistor in the collector circuit. V_{CC} indicates the bias voltage applied across the emitter-collector circuit. Since the emitter-base and collector-base junctions are in series with each other, the applied voltage V_{CC} will be divided between the junctions based on their individual resistances. The voltage measured across the transistor collector and emitter is called V_{CE}. The voltage measured across the transistor collector-base junction is called V_{CB}. Since the collector-base junction has a much greater resistance than the base-emitter junction, the collector-emitter voltage V_{CE} is larger than collector-base voltage V_{CB}. Base voltage V_{CB}, however, is greater than base-emitter voltage V_{BE}.

TRANSISTOR OPERATING CHARACTERISTIC CURVES

Transistors are manufactured to perform different functions under a variety of operating conditions. The transistor can be made to operate with different combinations of voltages and currents to obtain a certain performance. To obtain as much information as possible about a transistor, a set of operating characteristic curves can be used. The operating characteristic curves are provided by the manufacturer. See Figure 6-12.

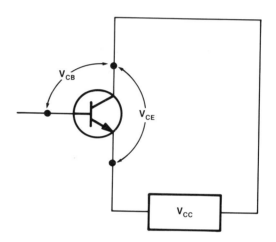

Figure 6-11. V_{CC} indicates the entire applied voltage from collector to emitter. V_{CB} indicates the voltage drop from collector to base. V_{CE} indicates the voltage drop from collector to emitter.

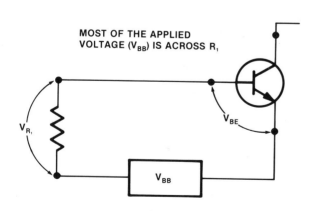

Figure 6-10. In the base circuit, V_{BB} indicates the entire applied voltage from base to emitter. V_{BE} indicates the voltage drop across the base-emitter. V_{R1} indicates the voltage drop across the series-limiting resistor R_1.

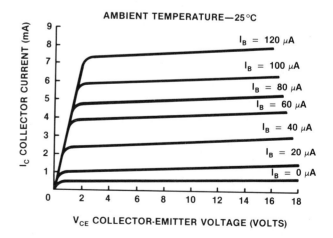

Figure 6-12. The collector characteristic curves indicate how the collector current I_C changes when there is a change in the base current I_B or collector-emitter voltage V_{CE}.

BASE TEST CIRCUIT   **COLLECTOR TEST CIRCUIT**

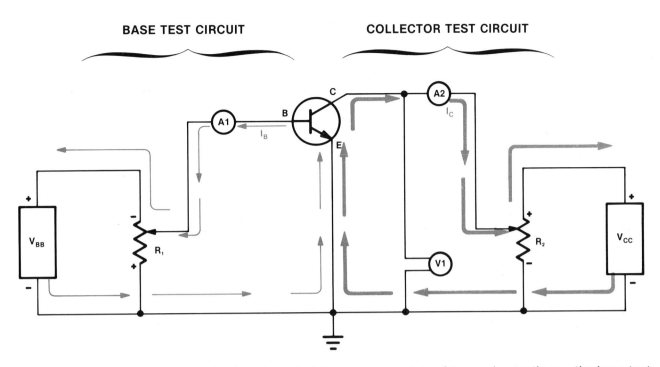

Figure 6-13. The test circuit for plotting characteristic curves consists of two major sections—the base test circuit and the collector test circuit.

The curves on this graph are known as collector characteristic curves. These curves indicate how collector current I_C changes when there is a change in the base current I_B or collector-emitter voltage V_{CE}.

NOTE: The vertical axis is labeled I_C for collector current and is measured in milliamps (mA). The horizontal axis is labeled V_{CE} for collector-emitter voltage and is measured in volts (V).

Before using the entire set of curves, it should be understood how a curve is generated. A test circuit can be used to create the collector characteristic curve. See Figure 6-13. The circuit is divided into two major sections—the base test circuit and the collector test circuit.

In the base test circuit, the voltage V_{BB} provides the forward bias for the base-emitter junction. Potentiometer R_1 is used to vary this bias to any specific value needed. Ammeter $A1$ is connected in series with base B. It is used to determine the value of I_B.

In the base-collector circuit, voltage V_{CC} provides reverse bias to the collector circuit. Potentiometer R_2 is used to vary this bias to any specific value needed. Ammeter $A2$ is connected in series with the collector C. It is used to determine the value of collector current I_C. To determine V_{CE}, voltmeter $V1$ is placed across the collector-emitter junction.

With this circuit, a single collector characteristic curve for the transistor can be plotted. See Figure 6-14. The vertical axis indicates I_C and is measured in

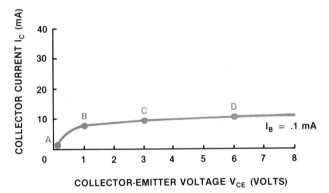

Figure 6-14. On the graph, point B represents a collector-emitter voltage of 1 volt and a collector current of 8 milliamps with the base held constant. Each point on the curve gives similar information about the operation of the transistor.

milliamps. The horizontal axis indicates collector-emitter bias V_{CE} and is measured in volts.

The first step in plotting the curve of Figure 6-14 is to adjust R_2 so that no voltage exists between the collector and emitter. This is done by moving the potentiometer to its lowest value or grounded condition. Next, R_1 is adjusted to produce a forward bias on the base-emitter circuit so that .1 milliamp (100 μA) is flowing in the base circuit. $A1$ is used to determine this value. At this point, with 1 milliamp base current and no collector voltage, collector current is zero. Point A represents a collector-emitter voltage of zero and a collector current of zero with the base held constant at .1 milliamp.

If R_2 is adjusted away from ground, a voltage would be applied between the collector-emitter. In this case, the potentiometer is adjusted to produce a V_{CE} of 1 volt. Since base current must remain the same, R_1 can be adjusted if necessary to maintain a base current of .1 milliamp. Depending upon the transistor being tested, a certain amount of base current will begin to flow. For this example, *A2* indicates 8 milliamps. This is shown on the graph. Point B represents a collector-emitter voltage of 1 volt and a collector current of 8 milliamps with the base held constant.

Next, R_2 is adjusted to produce a V_{CE} of 3 volts while the base circuit is still maintained at .1 milliamp. This increase in voltage causes the collector current to rise to 9 milliamps. In Figure 6-14, point C represents a collector-emitter voltage of 3 volts and a collector current of 9 milliamps with the base current held constant. If the procedure is repeated, R_2 is adjusted to a V_{CE} of 6 volts at a base current of .1 milliamp. The collector current will rise to 10 milliamps. Point D represents a V_{CE} of 6 volts and a collector current of 10 milliamps with the base current held constant.

If more information is needed on this transistor, additional curves can be plotted. See Figure 6-15. When R_1 is adjusted to produce more or less base current, it causes the curves to be above or below the original curve. The additional curves are denoted by the dotted lines.

All manufacturers produce collector characteristic curves for their devices. To make their curves very accurate, many test points are established and the curves are accurately drawn. Collector characteristic curves are essential in predicting the operating condition of a transistor in a circuit.

Obtaining Information from a Collector Characteristic Curve

With accurate curves, reasonable predictions can be made about the operation of a transistor. See Figure 6-16. For example, to determine the amount of collector current that would flow if the collector-emitter voltage V_{CE} is 4 volts and the base current is 0.4 milliamp, make a vertical line running from 4 volts until it crosses the 0.4 milliamp base current line. Mark this point. Draw another line horizontally across to the collector current axis. At this point, the collector current is approximately 33 milliamps. Thus, for a 4-volt V_{CE} and a constant base current of 0.4 milliamp, a 33-milliamp current in the collector circuit could be expected.

Beta (β) or h_{fe} to Describe Transistor Operation

When base current I_B is increased, collector current I_C will increase depending upon the voltage applied to the collector-emitter circuit. By using characteristic curves, it can be shown precisely how powerful the control of I_B is and how to predict its behavior. Technically, any increase in I_B that causes an increase in I_C is called the *current gain* of the transistor. Current gain in a transistor is designated by the Greek letter beta (β) or as h_{fe}. Both β and h_{fe} are used interchangeably.

NOTE: Designations beginning with h are used to describe operating characteristics for transistors. In the case of h_{fe}, it is used to designate current gain.

Beta or h_{fe} expresses a ratio. This ratio indicates a change in I_C compared to a change in I_B with the collector-emitter voltage V_{CE} constant:

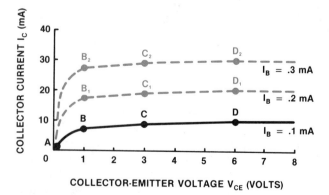

Figure 6-15. When additional information is needed on a transistor, additional curves are plotted (denoted by the dashed lines).

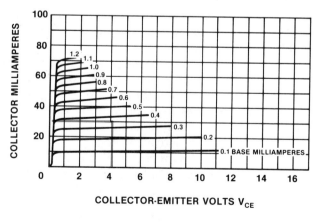

Figure 6-16. With accurate curves, predictions can be made about the operation of a transistor. In this case, it can be determined that for a 4-volt V_{CE} and a constant base current of 0.4 milliamp, a 33-milliamp current in the collector could be expected.

$$\beta \text{ or } h_{fe} = \frac{\Delta I_C}{\Delta I_B}$$

The Greek letter delta (Δ) means "a change in value."

Beta or h_{fe} is used to describe the operation of a transistor because current changes are more dramatic in a transistor than voltage changes. The input of a transistor is forward biased PN junction. Therefore, input voltages across this junction are small and difficult to measure. With small voltage changes, it is difficult to compare transistor input voltage with transistor output voltage. The PN junction does, however, draw substantial current flow for a small signal input voltage, and the current can be easily measured. Current flow in the base controls I_C in the output. Therefore, most transistors should be compared on the basis of their ratio of current input to current output rather than their voltage.

Characteristic Curves to Find β or h_fe

In the curves of Figure 6-17, a variety of base currents ranging from 0.1 milliamp to 1.2 milliamps are available. To determine what type of current change takes place in the collector circuit, it is necessary to find how much change takes place in the base current I_B. For example, if I_B changes from 0.1 milliamp to 0.7 milliamp, there is a change in I_B of 0.6 milliamp ($\Delta I_B = 0.6$ milliamp). At 0.1 milliamp (I_B), the collector current I_C is 10 milliamps, and at 0.7 milliamp (I_B), I_C is 50 milliamps. The change in I_C is 40 milliamps while the collector-emitter voltage remains constant at 3 volts. With the change in I_B known and the change in I_C known, β or h_{fe} can be calculated as follows:

$$h_{fe} \text{ or } \beta = \frac{\Delta I_C}{\Delta I_B} \quad \text{with } V_{CE} \text{ constant}$$

$$h_{fe} \text{ or } \beta = \frac{40 \text{ mA}}{0.6 \text{ mA}} \quad \text{with 3 volts constant}$$

$$h_{fe} \text{ or } \beta = 66$$

The number 66 designated as current gain means that any change in I_B causes a change in I_C 66 times greater. In terms of control, this device is extremely powerful. For each input, the output will be 66 times greater.

TRANSISTOR AS A DC SWITCH

The primary reason for the rapid development of the transistor was to replace mechanical switching. The transistor has no moving parts and can switch ON and OFF quickly.

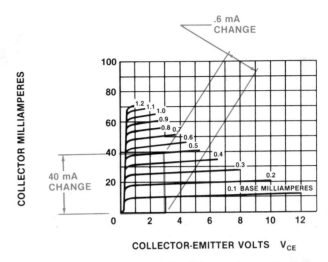

Figure 6-17. With only a .6-milliamp change in base current, a 40-milliamp change takes place in the collector circuit.

Mechanical switches have two conditions: open and closed or ON and OFF. The switch has a very high resistance when open and a very low resistance when closed. A transistor can be made to operate much like a switch. For example, it can be used to turn a pilot light PLI ON or OFF. See Figure 6-18. In this circuit, the resistance between the collector C and the emitter E is determined by the current flow between the base B and E. When no current flows between B and E, the collector-to-emitter resistance is high like that of an open switch. See Figure 6-18, left. The pilot light does not glow because there is no current flow.

If a small current does flow between B and E, the collector-to-emitter resistance is reduced to a very low value like that of a closed switch. See Figure 6-18, right. Therefore, the pilot light will be switched ON and begin to glow.

A transistor switched ON is usually operating in the *saturation region*. The saturation region is the maximum current that can flow in the transistor circuit. At saturation, the collector resistance is considered zero and the current is limited only by the resistance of the load. Mathematically it is expressed as:

$$I_{SAT} = \frac{V_{CC}}{R_L}$$

When the circuit in Figure 6-18 reaches saturation, the resistance of pilot light PLI is the only current-limiting factor in the circuit. When the transistor is switched OFF, it is operating in the *cutoff region*. The cutoff region is the point at which the transistor is turned OFF and no current flows. At cutoff, all the voltage is across the open switch (transistor) and the collector-to-emitter voltage V_{CE} is equal to the supply voltage V_{CC}. The characteristic

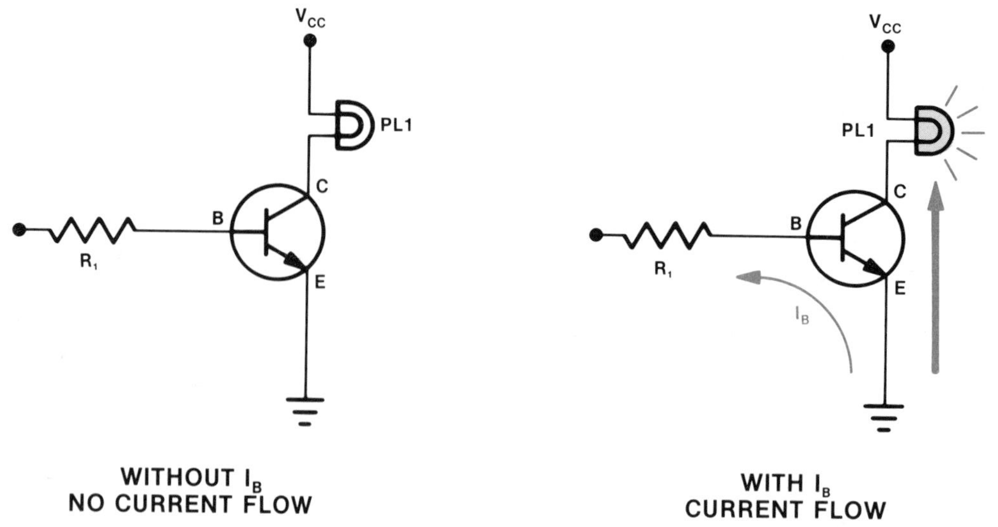

WITHOUT I_B
NO CURRENT FLOW

WITH I_B
CURRENT FLOW

Figure 6-18. With no emitter-base current I_B, the light remains OFF (left). With I_B flowing, the transistor is switched ON. It delivers current to the lamp, causing it to glow (right).

curve of Figure 6-19 shows the point of saturation (point A) and cutoff (point B).

ESTABLISHING A LOAD LINE

The *load line* is drawn between two extremes of operation. See Figure 6-19. It is assumed that the transistor is first saturated and then cut off. Point A saturation indicates the maximum current that can flow in the collector at that voltage. Saturation is the maximum output the transistor can have. At point B cutoff, there is no collector current and all the applied voltage source V_{CC} is dropped across the collector-emitter junction. At cutoff, the output of the transistor is completely shut down. Thus, the load line shows the maximum and minimum capabilities of a particular transistor. With this information, an operating point can be selected to give the outputs needed.

To calculate saturation for a load line, divide the load resistor (R_L) value into the voltage source V_{CC} to determine maximum current:

Saturation $\quad I_{SAT} = \dfrac{V_{CC}}{R_L}$

Example:
Given: $R_L = 2{,}150 \ \Omega$ and $V_{CC} = 15$ V

$$\frac{15 \ \text{V}}{2{,}150 \ \Omega} = .007 \ \text{A or } 7 \ \text{mA}$$

To find cutoff, select the desired V_{CC} (in this case, $V_{CC} = 15$ volts).

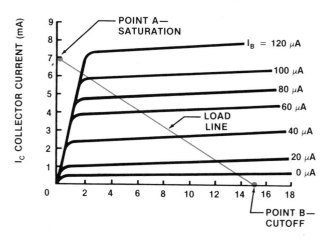

AMBIENT TEMPERATURE = 25°C

Figure 6-19. The saturation point A indicates the maximum current that can flow in the collector at that voltage. The cutoff point B indicates there is no collector current. All the applied voltage source V_{CC} is dropped across the collector-emitter junction.

BIASING TRANSISTORS

In order to achieve proper operation, the transistor must be properly biased. A transistor biased for saturation has the base-collector junction reverse biased. A transistor biased for cutoff has both junctions reverse biased.

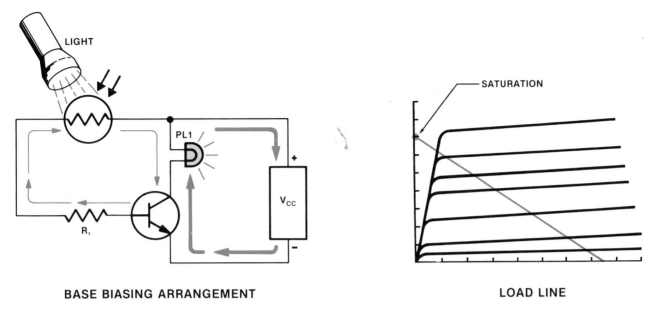

Figure 6-20. When the light shines on the photoconductive cell, the resistance of the photoconductive cell decreases and base current increases. Current in the collector circuit increases, and light turns ON. At this point, the transistor has reached saturation.

If a transistor is biased for cutoff, it can be turned ON again only by applying a forward-bias voltage (signal) to the emitter-base. However, the signal must be strong enough to overcome the reverse bias and produce a sufficient current flow for the transistor to reach saturation. When the signal is removed from the transistor, it returns to reverse bias and cutoff. Using this technique, the transistor can be switched OFF and ON rapidly.

Base Bias

Increasing forward-bias voltage on the base causes the current through the transistor to increase. When the forward-bias voltage on the base decreases, current through the transistor decreases. The circuit of Figure 6-20, left, illustrates the base biasing arrangement using a photoconductive cell as part of the base circuit. The load line for this circuit is shown in Figure 6-20, right.

Light shining on the photoconductive cell causes the resistance of the photoconductive cell to decrease. With reduced base resistance, the current into the base increases. This increase in current at the base-to-emitter junction causes the collector current to increase. The base current is limited only by the bias-limiting resistor R_1. If this bias value has been properly selected, the transistor will approach saturation and the pilot light turns ON.

Removal of light from the photoconductive cell causes the resistance of the photoconductive cell to increase. See Figure 6-21. With the increased resistance, the current into the base decreases. The decrease in current at the base-to-emitter junction causes the collector current to fall. If the bias value has been properly selected, the transistor will reach cutoff and the pilot light turns OFF.

To make the transistor more flexible, an electromechanical relay can be added to the output circuit. See Figure 6-22. With the addition of a relay with normally closed (NC) contacts, a photoconductive cell can be used to turn OFF a pilot light when light strikes the photoconductive cell. This type of circuit is often used in street light switching circuits.

NOTE: When a transistor is used to switch a relay, a diode should be connected across the relay coil. This diode keeps the reverse-inductive surge of current of the relay coil from damaging the transistor when the relay is switched OFF.

Fixed-Bias Configurations

Figure 6-22 shows how to properly bias a transistor from a single fixed-voltage source. The voltage source V_{CC} is a definite block or section in the circuit. This was done for illustrative purposes only. However, in most cases, the circuit will look like the circuit shown in Figure 6-23. Electrically, it is the same circuit as the circuit shown in Figure 6-22. However, the emitter E is connected to ground, which is the negative terminal of the voltage source. The collector and R_1 are connected to V_{CC} which represents the positive

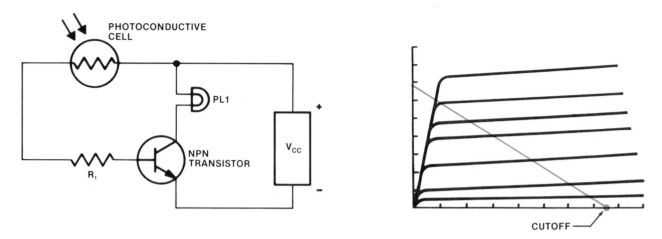

Figure 6-21. Removal of light from the photoconductive cell causes its resistance to increase. Current in the base decreases and current in the collector circuit decreases, and the light turns OFF. At this point, the transistor has reached cutoff.

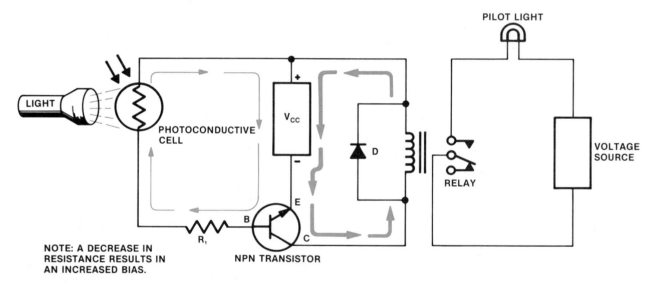

Figure 6-22. With the addition of a relay with normally closed (NC) contacts, a photoconductive cell can be used to turn OFF a lamp when light strikes the photoconductive cell.

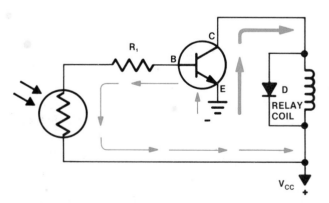

Figure 6-23. Although circuits may be drawn differently, electrically they are identical. In this circuit, the emitter *E* is connected to ground, which is the negative terminal of the voltage source.

terminal of the source. This arrangement is used because it is easier to represent a transistor circuit when several transistors are using power from a common source.

The same circuit may be shown in yet another way for convenience of layout. See Figure 6-24. The only difference is that the negative terminal of the voltage source is represented by the ground symbol, and the positive terminal of the source is represented by the positive sign and V_{CC} designation. Electrically, all the circuits (Figures 6-22, 6-23, and 6-24) are exactly the same.

Base Bias Instability

Base bias is one of the simplest methods of transistor biasing. With this type of bias, the amount of base

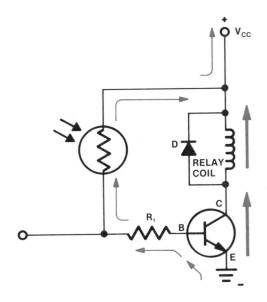

Figure 6-24. In this circuit, the negative is the ground and the positive is indicated by the sign V_{CC}.

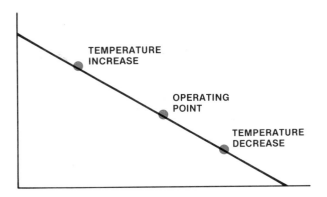

Figure 6-25. Any change in the ambient temperature causes the operating point to shift up or down the load line.

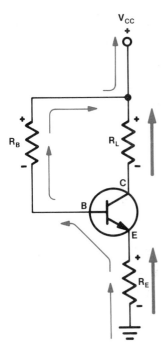

Figure 6-26. Emitter-feedback bias is accomplished by placing a resistor R_E in series with the emitter.

Bias Stabilization Circuits

A change in ambient temperature may cause thermal instability which, if uncontrolled, results in thermal runaway. Thermal runaway can destroy the transistor. In addition to heat sinking, transistors can be protected by bias stabilization circuits. Bias stabilization circuits are designed to counteract the cumulative current increase due to a rise in temperature. The most common bias stabilization circuits are *emitter-feedback*, *collector-feedback*, and *combination bias*. In addition, thermistors and diodes may be used as part of the circuit stabilization to provide thermal stabilization.

Emitter-Feedback Bias Stabilization Circuits. Figure 6-26 shows an emitter-feedback bias stabilization circuit. By placing resistor R_E in series with the emitter E, a voltage develops at E and opposes the bias voltage at the base. The resistance values of R_B and R_E must be chosen so that the proper base-emitter bias current flows under ordinary operating conditions. Then, if the transistor temperature changes to increase bias currents, the voltage drop across R_E increases to increase the emitter voltage. The increased emitter voltage opposes the input bias voltage and reduces the base current. The collector current is then brought back to normal.

Collector-Feedback Bias Stabilization Circuits. Collector feedback is also called *degenerative feedback*

current is dependent upon the value of the limiting resistor and the resistance of the emitter-base junction. As a signal reaches the limiting resistor, the signal will add to or subtract from the bias value to establish a variation in the output current of the device. Under normal stable temperatures, a base bias circuit will perform as it should.

However, if the ambient temperature changes, the junction resistance also changes and the bias current is affected. Any change in bias shifts the operating point on the load line. See Figure 6-25. A shift of the operating bias due to heat results in moving the operating point closer to saturation or cutoff. The change in bias due to heat is called *thermal instability*. Base bias is easily affected by thermal instability. *Bias stabilization circuits* must be used to protect transistor circuits from thermal instability.

or *negative feedback*. Collector feedback is accomplished by taking part of the output voltage and feeding it back into the input to return the circuit to normal operation. In a collector-feedback stabilization circuit, the input current-limiting resistor R_B is connected directly to the collector C rather than directly to the voltage source. See Figure 6-27. The bias voltage now consists of a base voltage minus the drop across R_B. Thus, if higher temperatures cause the transistor bias current to increase, the increased collector current causes a larger drop across the output resistor R_L. The voltage at C will then decrease. This will reduce the input bias current, causing the collector current to decrease and return to normal. The opposite happens if the transistor temperature decreases.

Combination Bias Stabilization Circuits. Both emitter feedback and collector feedback are often used to form a combination bias stabilization circuit. See Figure 6-28. The use of both bias voltages results in maintenance of good thermal stability.

Voltage Divider with Emitter Bias. The bias for a transistor is often set by a voltage divider network. See Figure 6-29. R_1 and R_2 form a voltage divider. They also set the fixed bias at the base terminal. Their resistance values are selected to provide the required bias current. Resistor R_E is the temperature stabilizing resistor that determines the voltage on the emitter. Emitter current I_E is equal to the sum of base and collector currents. Current flow through R_E develops a voltage drop that opposes the base bias. Thus, any increase in the voltage drop across R_E due to temperature changes opposes the input bias and reduces collector current.

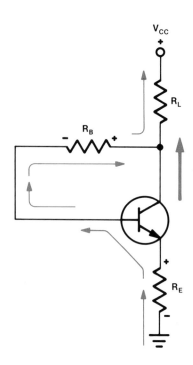

Figure 6-28. Combination bias requires the use of both emitter feedback and collector feedback.

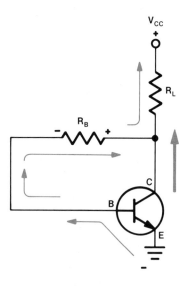

Figure 6-27. Collector feedback is accomplished by taking part of the output voltage and feeding it back into the input to return the circuit to normal operation.

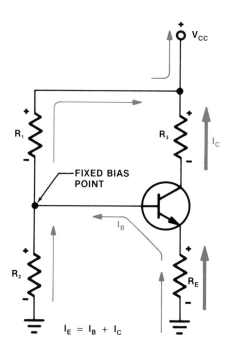

Figure 6-29. By establishing a voltage divider network, the voltage drop across R_E (due to temperature changes) opposes the input bias and reduces collector current.

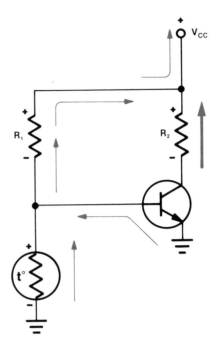

Figure 6-30. A thermistor can be used to form a voltage divider circuit. As ambient temperature changes, the thermistor changes in resistance, helping stabilize the circuit.

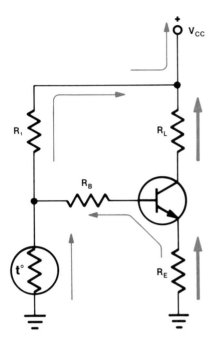

Figure 6-31. The thermistor can be used with other types of stabilization circuits to provide more effective control.

Thermistor Bias. Thermistors are also used for bias stabilization. Since a thermistor changes resistance with a change in temperature, it can be used as part of a stabilization circuit. The thermistor forms a voltage divider with R_1 to provide a fixed bias for the base. See Figure 6-30. As temperature increases with an increase in base bias current, the resistance of the thermistor decreases. With less resistance, the thermistor drops less voltage for the base, and the base bias decreases. With a properly designed circuit, the thermistor will change the base current to compensate for changes in temperature.

Thermistor/Emitter Feedback Bias. A thermistor can be used as part of a voltage divider with an emitter-base resistor R_E. See Figure 6-31. The base is supplied with fixed bias through resistor R_B. R_E provides normal stabilization. By adding the thermistor, however, the current stabilization process becomes more effective.

The circuit operates as follows: When the temperature increases, there is a tendency for the base current to increase. The increase in base current causes the resistance of the thermistor to decrease. The increase changes the bias on the base. The change in bias causes the base current to decrease by the same amount it increased, thus keeping the base current constant. The advantage of using the thermistor is that stabilization tends to occur before bias currents can change.

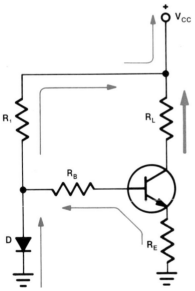

Figure 6-32. By using a diode instead of a thermistor, the operating characteristics of a transistor can be more closely matched.

Diode Stabilization. Matching a thermistor to the operating characteristic of a transistor is often a problem with thermistor stabilization. A diode is often used instead of a thermistor since its PN junction closely approximates that of a transistor. When a diode is used, it is forward biased. In stabilizing the circuit, the diode causes the same current changes as the thermistor does. See Figure 6-32.

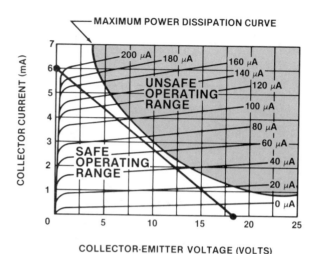

Figure 6-33. The maximum power dissipation curve shows the safe operating range of a transistor. Under no circumstance should the operating point be at the right of the maximum power dissipation curve.

POWER DISSIPATION

Because a transistor is extremely small, it can dissipate only a certain amount of heat (power dissipation factor) before it begins to change operating characteristics. If an overload continues for even a short period of time, the transistor junction will be destroyed by thermal runaway. Transistors are rated for a certain wattage at 25 °C. If the temperature drops below 25 °C, more power can be handled. If the temperature exceeds 25 °C, the transistor must be derated (reduced current through transistor) or a heat sink must be added to dissipate the additional heat. To determine the wattage output at any given point, characteristic curves can be used. To determine the amount of wattage, multiply the collector voltage V_{CE} times the collector current I_C ($P = EI$). If the V_{CE} is 10 volts and I_C is 10 milliamps (.010 amps), then $P = 10$ volts $\times$ 10 milliamps, or 100 milliwatts.

Maximum Power Dissipation Curves

In order to protect the transistor, the maximum power dissipation curve must be considered. See Figure 6-33. It represents the maximum power that a transistor can handle under normal ambient temperatures.

The maximum power disspation curve shows the safe operating ranges of a transistor. The shaded areas in Figure 6-33 indicate the areas in which the transistor will have problems. Under no circumstances should the load line move to the right of the maximum power dissipation curve. It should stay to the left of the curve. Circuit designers usually try to have operation well within the maximum power dissipation curve to help

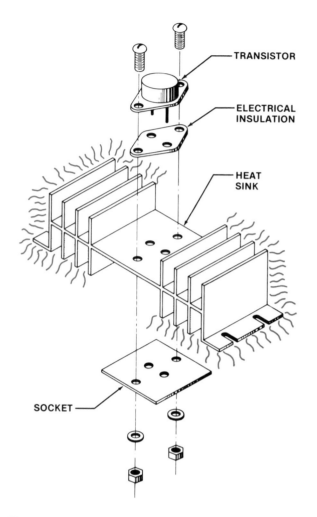

Figure 6-34. Transistors, like diodes, use heat sinks to dissipate excess heat and provide thermal protection.

compensate for temperature variations. This compensates for situations in which there is no control of temperatures—for example, in rooms without air conditioning in summer months.

Transistors/Heat Sinks

Transistors, like diodes, use heat sinks for thermal protection. The cases of certain power transistors are designed specifically for ease in cooling. Some power transistors use radial fins (part of the transistor design) for conducting heat away. Other power transistors are designed for used with heat sinks. See Figure 6-34. In some cases, the chassis of the equipment is used as a heat sink. See Figure 6-35.

When transistors are required to operate in high ambient temperatures, forced cooling by a fan or by air conditioning is used. In some precision solid state equipment, the unit will actually shut down if ambient

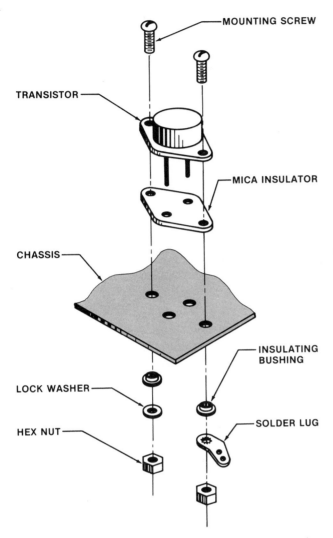

Figure 6-35. In some cases, the chassis is used as a heat sink.

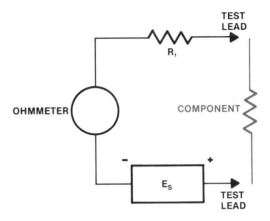

Figure 6-36. The ohmmeter is a low-current ammeter placed in series with a voltage source E_S and current-limiting resistor R_1.

temperatures exceed safe levels. Another cause for shutdown is if the forced cooling system is shut down for some reason.

TESTING TRANSISTORS

Although there are a variety of transistor testers available, an ohmmeter provides much of the information necessary to determine the proper operation of a transistor.

Testing Transistors for NPN and PNP

Determining whether a transistor is an NPN or a PNP is impossible just by visual inspection. An ohmmeter must be used to identify the transistor type. The ohmmeter is a low-current ammeter placed in series with a voltage source E_S and a current-limiting resistor R_1. See Figure 6-36. A component placed

between the test leads completes the series circuit. It also determines the amount of current flow through the meter. (The test leads have a definite polarity which is determined by the internal voltage source.) The polarity of the leads are used to forward and reverse bias the PN junction. For example, if the positive probe is placed on the P-type material and the negative probe is placed on the N-type material, a diode will be forward biased. The forward-biased diode indicates current flow and a low resistance. See Figure 6-37. When the leads are reversed, the PN junction will be reverse biased. This indicates a high resistance and no current flow. See Figure 6-38. The polarity of the ohmmeter can be used to determine which leads are connected to P-type material and N-type material in NPN and PNP transistors.

NOTE: Red and black test leads on the outside of the ohmmeter do not necessarily indicate which lead is positive and which lead is negative. Check the manufacturer's operating manual.

Once the ohmmeter lead polarity is known, the following procedure is used to differentiate between NPN and PNP transistors:

1. Determine which transistor leads are the emitter (*E*), base (*B*), and collector (*C*).

2. For Test 1, place the positive ohmmeter lead on *B* and the negative lead on *E*. See Figure 6-39.

3. Record the resistance.

4. For Test 2, reverse the leads on *B* and *E*. See Figure 6-40.

5. Record the resistance.

One test should indicate that the junction has a low resistance and is forward biased. The other test should indicate that the junction has a high resistance and is reverse biased. If Tests 1 and 2, with the polarities

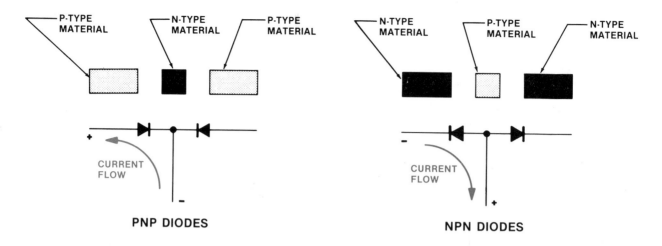

Figure 6-37. When the PN junction is forward biased, current flows and there is a low resistance reading.

shown, indicate these readings, the transistor is a PNP. If the readings are reversed, it is an NPN transistor.

Testing Transistors for Opens and Shorts

High voltages, improper connections, and overheating can damage a transistor. The technician is responsible for determining the condition of the transistor. Although many types of transistor testers are available, the technician may not have one when it is needed. Therefore, an ohmmeter can be substituted for a transistor tester when finding opens and shorts. The transistor may be considered back-to-back diodes when conducting tests for opens and shorts. See Figure 6-41.

NOTE: A diode passes current when it is forward biased. It blocks current when it is reverse biased.

When making resistance measurements, the ohmmeter should indicate a low resistance one way and a high resistance the opposite way.

The first diode junction to test is the emitter-to-base junction. See Figure 6-42. Placing the leads one way and then reversing the leads should result in a high forward-to-reverse resistance ratio. The second diode junction of the base-collector should show a similar high forward-to-reverse resistance ratio. See Figure 6-43. The last check is from emitter to collector. See Figure 6-44. It should show a high resistance in both directions for the transistor to be considered good.

A comparison should be made of the resistance measurements of the emitter-base section in forward and reverse. The measurements should indicate a high resistance in one direction and a low resistance in the

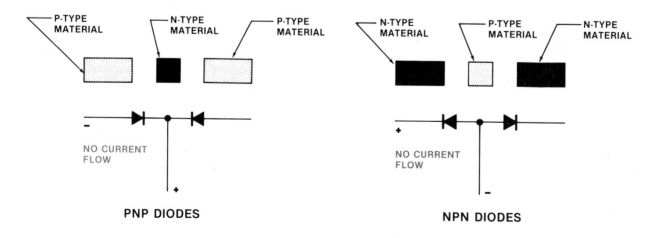

Figure 6-38. When the PN junction is reverse biased, there is a high resistance reading and no current flows.

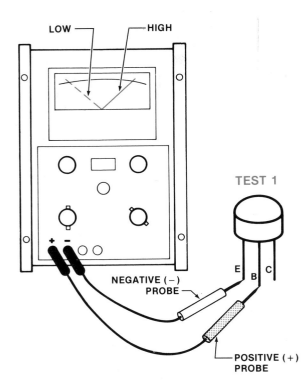

Figure 6-39. Depending upon the type of transistor (PNP or NPN), the ohmmeter will have either a high or low resistance reading in Test 1.

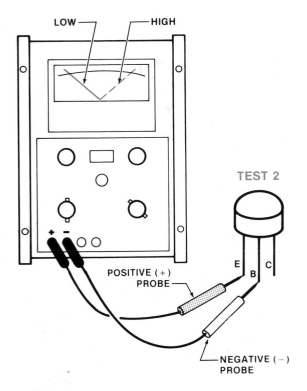

Figure 6-40. When the polarity on the transistor is reversed, the ohmmeter produces either a high or low resistance reading which is opposite that of Test 1 in Figure 6-39.

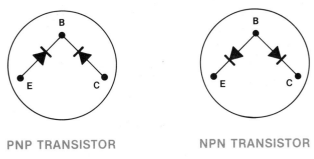

PNP TRANSISTOR   NPN TRANSISTOR

Figure 6-41. The PN junctions of a transistor can be considered back-to-back diodes when testing for opens and shorts.

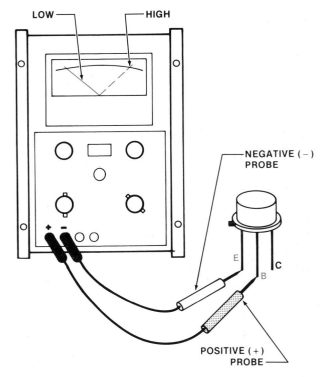

Figure 6-42. Placing the ohmmeter leads on the emitter-base one way then reversing them should result in a high forward-to-reverse resistance ratio.

opposite direction if the transistor is good. If both emitter-base resistances are low, the transistor is shorted. If both resistances are high, the transistor is open.

The measurements of the base-collector terminals should indicate a high resistance in one direction and a low resistance in the opposite direction if the transistor is good. If both base-collector resistances are low, the transistor is shorted. If both resistances are high, the transistor is open. If the resistance from emitter to collector is low in either direction, the transistor is shorted.

When comparing the measurements of the emitter-collector terminals, the measurements should show a high resistance in both directions. If the resistance is low in either direction, the transistor is shorted.

Testing Transistor Switches

It is impossible to determine whether a transistor is switched ON or OFF just by visual inspection. However, if an indicator light is in the collector circuit, its ON/OFF status can be determined. Usually, a voltmeter is used to determine the status of a transistor. Figure 6-45 illustrates the method for checking a transistor switch:

1. When the transistor switch is closed (ON), V_{CE} reads near 0 volts. A transistor in this condition has virtually no resistance just as an electromechanical switch would have no resistance. Therefore, the transistor has no voltage drop.

2. When the switch is open (OFF), V_{CE} will approximately equal V_{CC}. The particular V_{CC} reading in the open condition occurs because the load R_L has no current flowing through it. Approximately all the voltage is dropped across the collector-to-emitter.

Additional transistor checks are:

3. If the load fails to energize, check the voltage input to see that it is high enough to turn the transistor ON. If the voltage input is correct, a silicon transistor will have a V_{BE} reading of about 0.6 volt when turned ON. See Figure 6-46. The remaining input voltage will be dropped across the current-limiting resistor R_B.

4. If the voltage input is correct but voltage V_{BE} is considerably higher or lower than 0.6 volt, the base-emitter junction is bad. The transistor must then be replaced.

5. If V_{BE} is correct and V_{CE} is close to (approximately equal to) V_{CC}, the transistor has an open collector-base junction. The transistor must then be replaced.

6. If the transistor fails to turn OFF, check for a short circuit from collector to base or from collector to emitter. In either case, the transistor must be replaced if there is a short.

7. If V_{CE} is near 0 volts but the load does not turn ON, the load probably has an open circuit and must be replaced.

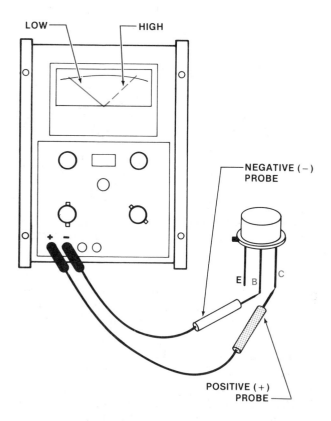

Figure 6-43. Placing ohmmeter leads on the base-collector one way then reversing them should also result in a high forward-to-reverse resistance ratio.

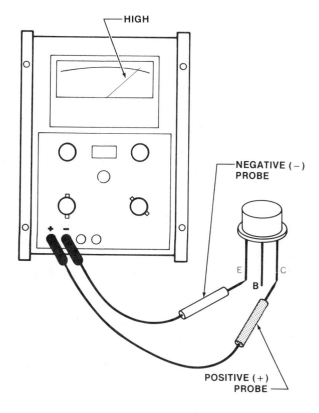

Figure 6-44. When comparing the measurements of the emitter-collector terminals, the measurements should show a high resistance in both directions.

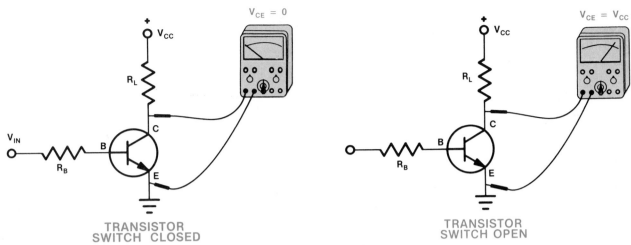

Figure 6-45. A voltmeter can be used to check the operating condition of a transistor. In this case, the emitter-collector voltage is used to determine if the transistor is ON or OFF.

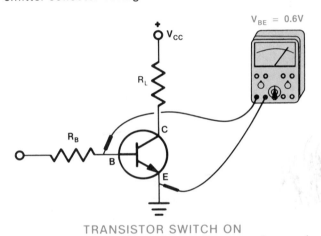

Figure 6-46. The base-emitter voltage can be used to determine if a transistor is ON or OFF.

TRANSISTOR SWITCHING APPLICATIONS

Transistors as single devices, or in multiples, are used for switching because of their reliability and speed. In certain situations, transistors are also integrated with other solid state components to form more complex devices. In each application, however, the fundamental operating principle of the transistor remains the same.

Seven-Segment Display Readout

By switching various combinations of transistors ON or OFF, different numbers can be created on the *seven-segment display readout*. See Figure 6-47. For example, if all the transistors (A through G) are switched ON, an 8 will appear on the display. If all are switched on except E and D, a 9 will appear. There is usually additional circuitry prior to the seven-segment transistor devices to help in decoding the proper signals to the display. When all circuitry is

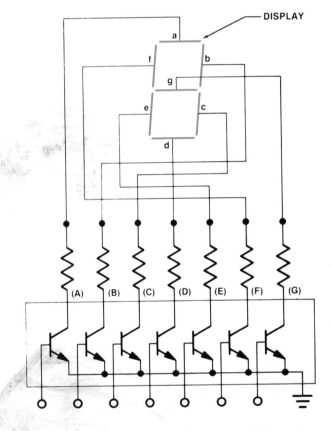

Figure 6-47. By switching ON or OFF various combinations of transistors, different numbers appear on the seven-segment display readout.

present, it is called as seven-segment decoder/driver display, or readout device.

Electronic Timer

An electronic circuit like the one shown in Figure 6-48 is used to time control a machine or process. The

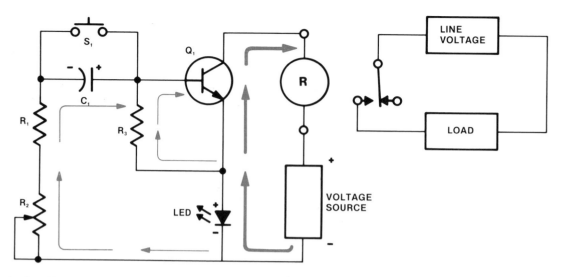

Figure 6-48. The electronic timer consists of a transistor and an *RC* time constant controlling a relay.

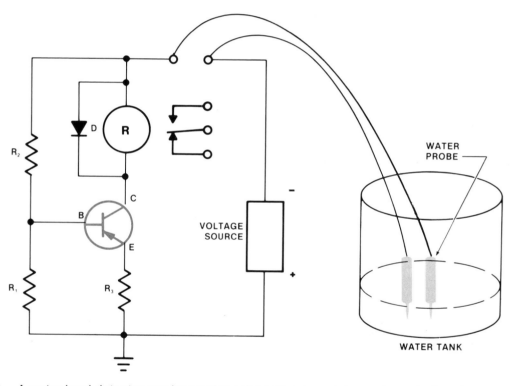

Figure 6-49. A water level detector can be provided through a transistor and a set of probes.

circuit uses the characteristic of an *RC* time constant to provide timing. This is developed by resistors R_1 and R_2 plus capacitor C_1. When the circuit is triggered by momentarily closing S_1, the NPN transistor Q_1 starts to conduct. Q_1 remains in conduction until the charge on C_1 increases to approximately the same voltage as the voltage source, and the transistor is cut off. If the time interval is to be changed, it is accomplished through the use of potentiometer resistance R_2.

If the resistance is increased, the time interval is increased. If the resistance is decreased, the time interval is decreased. The LED (light emitting diode) indicates when the load is ON.

Transistor Water Level Detector

The circuit of Figure 6-49 is a simple, low-cost transistor water level detector. The advantage of this circuit is its ability to use a very small amount of current (in the milliamp range) to control a much larger output such as a sump pump motor.

In this circuit, the amount of collector current determines when the relay will trip. The base is held negative with respect to the emitter by the voltage divider. The voltage divider consists of resistors R_1 and R_2. Current stabilization is maintained by resistor R_3.

As the water reaches a predetermined level set by the mounting height of the water probes, the bias circuit is complete. With the bias circuit complete, voltage will be applied to the circuit, causing base current to flow. With a base current established, the collector current will flow through the relay. If bias current has been properly established, the relay should trip and activate the circuit. Removing the water probes from the water causes the relay to de-energize.

Chapter 6 - Review Questions

1. What are the two types of transistors?
2. What are the three main parts of a transistor?
3. How are transistors referenced?
4. What is the bias arrangement for the normal operation of any transistor?
5. What is the critical factor in determining current flow through a transistor?
6. Define the following abbreviations: V_{BB}, V_{BE}, V_{CE}, V_{CB}.
7. What is used to determine the different voltage and current combinations possible in a transistor?
8. Explain beta.
9. How is beta determined mathematically?
10. In what region is a transistor operating when it is ON and when it is OFF?
11. What line is drawn between the two extremes of operation on a characteristic curve?
12. What should be placed across the relay coil of a relay that is used with a transistor?
13. Explain thermal instability.
14. What are the three most common types of base bias stabilization circuits?
15. What is the maximum power dissipation curve?
16. At what temperature is the power of a transistor rated?
17. In transistor operation, what is the function of a heat sink?
18. How is a transistor tested to determine if it is an NPN or a PNP?
19. How is it determined if a transistor junction is open?
20. What is the V_{CE} of a transistor that is switched OFF?

7 SILICON CONTROLLED RECTIFIER (SCR)

The *silicon controlled rectifier (SCR)* is a four-layer (PNPN) semiconductor device. It uses three electrodes for normal operation. See Figure 7-1. The three electrodes are the anode, cathode, and *gate*. The anode and cathode of the SCR are similar to the anode and cathode of an ordinary semiconductor diode. The gate serves as the control point for the SCR.

The SCR differs from the ordinary semiconductor diode in that it will not pass significant current, even when forward biased, unless the anode voltage equals or exceeds the *forward breakover voltage*. However, when forward breakover voltage is reached, the SCR switches ON and becomes highly conductive. The SCR is unique because the gate current is used to reduce the level of breakover voltage necessary for the SCR to conduct or fire.

There are many case styles for the SCR. See Figure 7-2. Low-current SCRs can operate with an anode current of less than one amp. High-current SCRs can handle load currents in the hundreds of amperes. The size of an SCR increases with an increase in its current rating.

Key Words

Forward blocking current	Half-wave phase control circuit	Phase control
Forward breakover voltage	Holding current	Silicon controlled rectifier
Gate	Hybrid circuit	

SCR CHARACTERISTIC CURVE

Figure 7-3 shows the voltage-current characteristic curve of an SCR when the gate is not connected. In reverse bias, the SCR operates like a regular semiconductor diode. With reverse bias, there is a small current until avalanche is reached. After avalanche is reached, the current increases dramatically. This current can cause damage if thermal runaway begins.

When the SCR is forward biased, there is also a small forward leakage current called the *forward blocking current*. This current stays relatively constant until the forward breakover voltage is reached. At that point, the current increases rapidly and is often called the forward avalanche region. In the forward avalanche region, the resistance of the SCR is very small. The SCR acts much like a closed switch and the current is limited only by the external load resistance. A short in the load circuit of an SCR can destroy the SCR if overload protection is not adequate.

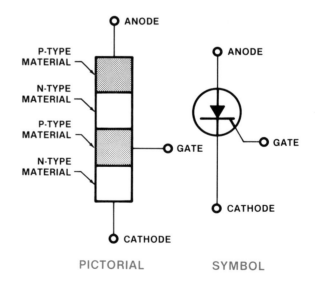

Figure 7-1. The silicon controlled rectifier (SCR) is a four-layer semiconductor device. Its three electrodes are the anode, cathode, and gate.

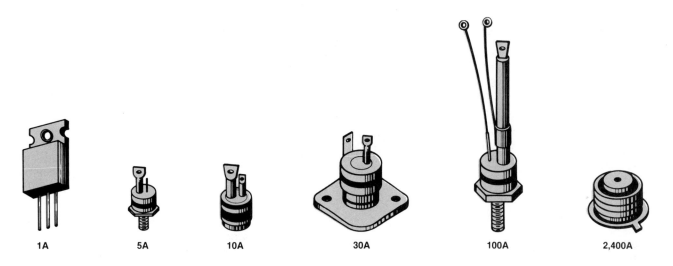

Figure 7-2. The SCR comes in a variety of case styles. Current ratings run from less than one ampere to several hundred amperes.

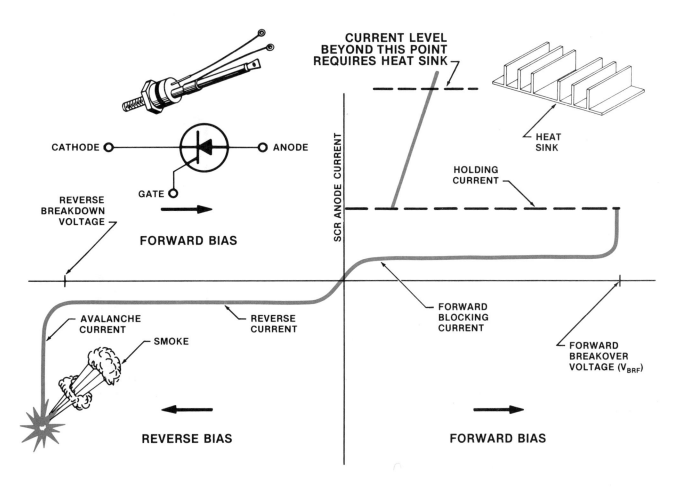

Figure 7-3. The voltage-current characteristic curve of an SCR shows that the SCR operates much like a regular diode in reverse bias. In forward bias, a certain value of forward breakover voltage must be reached before the SCR will conduct.

Operating States of an SCR

The SCR operates much like a mechanical switch. It is either ON or OFF. When the applied voltage is below the forward breakover voltage (V_{BRF}), the SCR fires, or is ON. The SCR will remain ON as long as the current stays above a certain value called the *holding current*. When voltage across the SCR drops to a value too low to maintain the holding current, it will return to its OFF state.

Gate Control of Forward Breakover Voltage

When the gate is forward biased and current begins to flow in the gate-cathode junction, the value of forward breakover voltage can be reduced. Increasing values of forward bias can be used to reduce the amount of forward breakover voltage (V_{BRF}) necessary to get the SCR to conduct. See Figure 7-4.

Once the SCR has been turned ON by the gate current, the gate current loses control of the SCR forward current. Even if the gate current is completely removed, the SCR will remain ON until the anode voltage has been removed. The SCR will also remain ON until the anode voltage has been significantly reduced to a level where the current is not large enough to maintain the proper level of holding current.

SCR CONSTRUCTION

The internal structure of an SCR consists of two PN junctions joined together to form a PNPN layer. See Figure 7-5. When reverse-bias voltage is applied, there is no current flow. Reverse bias exists when the positive lead of the voltage source is attached to the cathode and the negative lead is attached to the anode. The positive terminal of the voltage source at the cathode attracts electrons away from the PN junction nearest to the cathode. This creates a wide depletion region at that junction. The negative terminal of the voltage source at the anode attracts holes away from the PN junction nearest to the anode. Another wide depletion region is created at this junction. No appreciable amount of current flow takes place in the reverse-bias condition.

When forward bias is applied to the junctions, no appreciable amount of current will flow either. See Figure 7-6. In this case, the negative terminal (cathode) repels electrons through the N-type material toward the PN junction closest to the cathode. This in turn attracts holes from the P-type material toward the PN junction closest to the cathode and away from the center PN junction. The positive terminal repels holes through the P-type material toward the PN junction nearest to the anode and away from the PN junction in the center. The net result is the creation of a larger

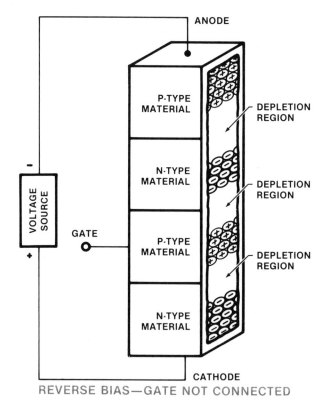

Figure 7-5. The internal structure of an SCR consists of two PN junctions joined together to form a PNPN layer. With reverse bias voltage applied to this layered structure, there is no noticeable current flow due to the formation of depletion regions.

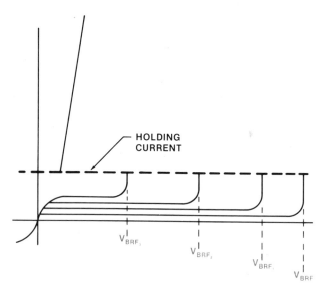

Figure 7-4. When the gate is forward biased and current begins to flow in the gate-cathode junction, the value of the forward breakover voltage can be reduced.

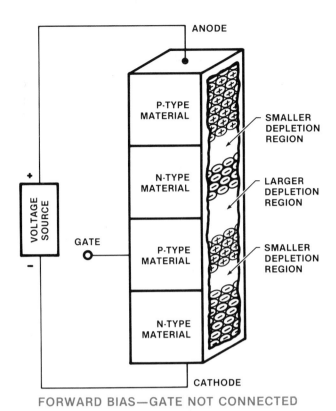

FORWARD BIAS—GATE NOT CONNECTED

Figure 7-6. With forward bias voltage applied to the layered structure, there is no noticeable current flow due to the formation of the larger depletion region in the center.

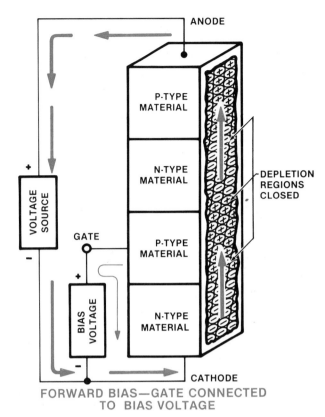

FORWARD BIAS—GATE CONNECTED
TO BIAS VOLTAGE

Figure 7-7. When the gate of the SCR is pulsed or triggered with a forward bias applied between the cathode and anode, the SCR will conduct. This current continues to flow as long as current is supplied. The SCR turns OFF when the current is turned OFF or when the forward breakover voltage is reduced or removed. Another gate pulse is required to restore conduction.

depletion region at the center PN junction. Therefore, there is very little current flow.

The SCR conducts in forward bias if the forward breakover voltage is reached or if a small forward bias voltage is applied to the gate. Figure 7-7 illustrates what happens internally when the gate is pulsed or triggered and the proper forward bias is applied to the SCR.

The positive terminal (anode) repels the holes in the P-type material toward the center junction closest to the anode. This attracts an equal number of electrons from the N-type material on the other side of the junction resulting in the depletion region closing at the center junction. All three junctions now have closed depletion regions, and the SCR will conduct current from the cathode to anode. This current will continue to flow as long as current is being supplied. The SCR turns OFF whenever the current is turned OFF or when the forward breakover voltage is reduced or removed. Another gate pulse is required to restore conduction.

SCR Equivalent Circuit

Another way to understand a complex device is to compare it to a simple device. This procedure is called the equivalent circuit analogy or comparison technique. The equivalent circuit for an SCR is a circuit consisting of an interconnected PNP transistor and an NPN transistor. See Figure 7-8. The emitters (*E*) of the PNP and NPN transistors represent the anode and cathode respectively. The base (*B*) and collector (*C*) of each transistor represent the gate connection. See Figure 7-9. In biasing the equivalent circuit (just as an SCR is biased), the anode is made positive and the cathode is made negative with no gate voltage applied. In this state, the NPN transistor will not conduct because its emitter-base junction is not forward biased. Since the NPN transistor has no base current, the emitter-base junction of the PNP transistor cannot conduct. Both transistors are turned OFF, therefore, no current flows. Without a gate potential, the SCR equivalent circuit, like the SCR, cannot conduct in either direction regardless of the polarity of the applied voltage.

To have the SCR turned ON, the gate must be connected and a voltage must be applied. See Figure 7-10. This can be done by applying a positive potential

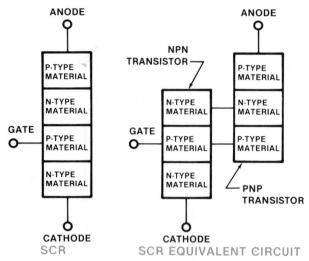

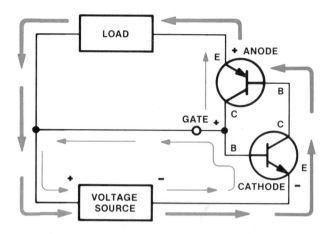

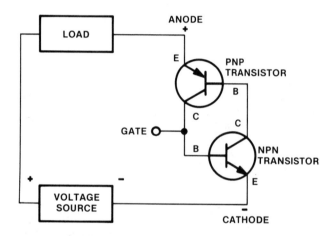

Figure 7-8. The equivalent circuit of an SCR can be constructed by interconnecting a PNP transistor and an NPN transistor.

Figure 7-9. In an equivalent SCR circuit, the emitters (*E*) of the PNP and NPN transistors represent the anode and cathode. The base (*B*) and collector (*C*) of each transistor is interconnected to represent the gate connection. Since no gate potential is applied, the SCR equivalent circuit, like an SCR, cannot conduct in either direction no matter what the polarity of the applied voltage is.

Figure 7-10. The SCR can start operating with only a momentary positive pulse to the gate. The pulse causes the emitter-base junction of the NPN transistor to be forward biased, which in turn causes base current to flow through the PNP transistor. Each transistor holds the other in a state of constant conduction. To switch the SCR circuit OFF, reduce the anode-to-cathode voltage to nearly zero, or momentarily open the circuit.

with respect to the cathode across the gate lead. With only a momentary pulse (as little as 500 μ seconds) to the gate, the circuit will start operating. The momentary pulse causes the emitter-base junction of the NPN transistor to be forward biased and to conduct. The current from the NPN transistor in turn causes base current to flow through the PNP transistor.

The collector current through the PNP transistor supports the base current through the NPN transistor.

The transistors hold each other in a state of constant conduction. The circuit was put into a state of conduction by only a momentary pulse to the gate. Once in conduction, the circuit is self-sustaining and the gate signal can be removed. In order to switch the SCR circuit OFF, the anode-to-cathode voltage must be reduced to nearly zero, or the circuit must be momentarily opened.

SCR as a DC Switch

The most basic application of an SCR is as a DC switch. An SCR used as a DC switch has many advantages over mechanical DC switching. The SCR provides arcless switching, low forward voltage drop, and a rapid switching time. It also lacks moving parts.

In the circuit of Figure 7-11, a single SCR is used to turn power ON and OFF to the load. The circuit is turned ON by momentarily closing the gate control switch S_1. Closing S_1 causes a very small gate current to flow through the current-limiting resistor R_1, which in turn causes the SCR to conduct. Once the SCR is ON, the gate signal can be removed and the SCR will remain in conduction.

In order to turn the SCR OFF, switch S_2 must be momentarily closed and switch S_1 must be open. See Figure 7-12. This shorts out the SCR for an instant and reduces the anode-to-cathode voltage to zero. With zero voltage, the forward current through the SCR drops below the holding value, and the SCR turns OFF.

The drawback to the circuit of Figure 7-12 is that switch S_2 must be closed and reopened across the high DC load current. Using a mechanical switch across an electronic switch defeats the purpose of electronic switching. This disadvantage can be overcome by using two SCRs and some additional circuitry. In the circuit of Figure 7-13, SCR_1 controls the DC power applied to the load. SCR_2, along with capacitor C and resistor R_1, turn SCR_2 and the load OFF.

The circuit of Figure 7-13 would operate as follows: When switch S_1 is momentarily closed, gate current flows through SCR_1 and resistor R_2, turning SCR_1 ON. This in turn allows current to flow to the load. Turning ON SCR_1 also grounds the left side of capacitor C to a negative potential through SCR_1. With one side of C grounded, C can charge through resistor R_1.

To turn SCR_1 OFF, switch S_2 must be momentarily closed with S_1 open, energizing the gate of SCR_2. See Figure 7-14. Gate current now passes through SCR_2 and R_2, causing SCR_2 to turn ON.

With SCR_2 conducting, the right side of C is now brought to ground potential placing the capacitor across SCR_1. The polarity of the voltage across the capacitor will then reverse bias SCR_1, dropping its forward current below the holding level (current). SCR_1 is shown in Figure 7-15.

NOTE: The capacitor is actually across SCR_1 with a completely opposite polarity.

Even if this reverse polarity takes place for only a moment, SCR_1 will stop conducting. To restart the circuit, S_1 would again have to be momentarily closed.

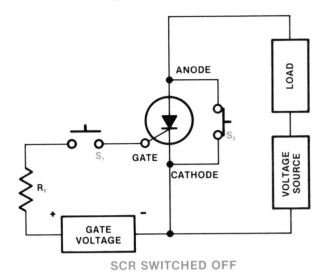

SCR SWITCHED OFF

Figure 7-12. Closing switch S_2 momentarily will turn OFF the SCR, provided that switch S_1 is open. Closing S_2 momentarily shorts out the SCR and reduces the anode-to-cathode voltage to zero. With zero voltage, the forward current drops below the holding value and the SCR turns OFF.

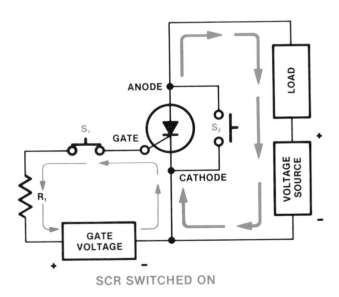

SCR SWITCHED ON

Figure 7-11. When switch S_1 is momentarily closed, a gate current is created through current-limiting resistor R_1, causing the SCR to conduct. Once the SCR is ON, the gate signal can be removed and the SCR will remain in conduction.

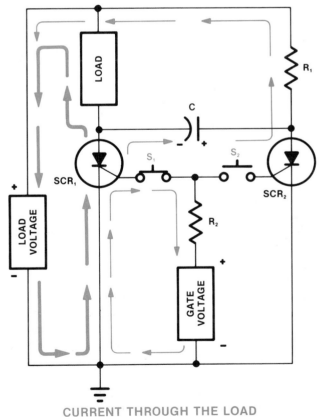

CURRENT THROUGH THE LOAD

Figure 7-13. When switch S_1 is closed, the gate current flows through SCR_1 and resistor R_2, turning SCR_1 ON. This in turn allows current to flow to the load. Turning ON SCR_1 also grounds the left side of capacitor C so that it may charge through R_1.

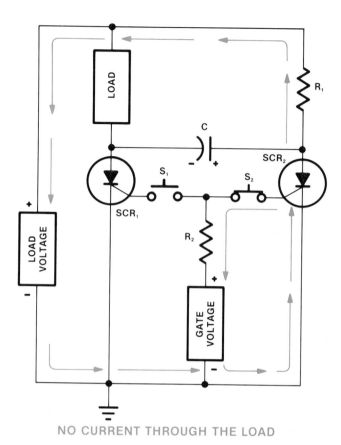

NO CURRENT THROUGH THE LOAD

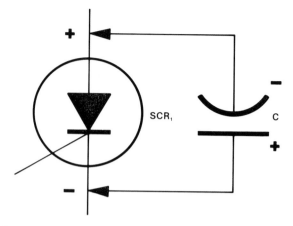

Figure 7-15. When a charged capacitor of opposite polarity is applied across an SCR, the SCR will stop conducting. The polarity of the voltage across the capacitor reverse biases the SCR so that the forward current of the SCR drops below the holding level and stops conducting.

NOTE: The entire cycle will continue producing full wave rectification indefinitely until S_1 and S_2 are opened. Opening S_1 and S_2 breaks the conducting paths to both gate circuits.

SCR FOR PHASE CONTROL

Regular diodes cannot be used for AC and DC switching while SCRs can. A regular diode is either completely ON or completely OFF. An SCR, on the other hand, can be turned ON at different points in the conducting cycle.

The ability of an SCR to turn ON at different points in the conducting cycle can be useful to vary the amount of power delivered to a load. This type of variable control is called *phase control*. The term phase control is used because phase refers to the time relationship between two events. In this case, it is the time relationship between the trigger pulse and the point in the conducting cycle when the pulse occurs. With phase control, the speed of a motor, the brightness of a lamp, or the output of an electric-resistance heating unit can be controlled.

The most basic control circuit using these principles is called a *half-wave phase control circuit*. See Figure 7-18. In this circuit, the input voltage is a standard 60-hertz line voltage. The illustration shows how this circuit would respond to the negative alternation of the 60-hertz line voltage.

When the negative alternation is applied to the circuit, the SCR will be reverse biased and will not conduct. In addition, diode D_1 will be reverse biased and no gate current will flow. Diode D_2 on the other

Figure 7-14. With switch S_2 momentarily closed, gate current will pass through SCR_2 and resistor R_2, causing SCR_2 to turn ON. With SCR_2 conducting, the right side of capacitor C is brought to ground potential, placing the capacitor across SCR_1. The polarity of the voltage across the capacitor will then reverse bias SCR_1, dropping its forward current below the holding level (current). To restart this circuit, switch S_1 must again be momentarily closed.

SCR as an AC Switch

Two SCRs are required for AC switching. See Figure 7-16. With S_1 closed, the gate will be connected for SCR_1 through diode D_1 and resistor R_1. During the positive half of each AC cycle, SCR_1 will be forward biased. With this forward bias, the gate current will flow through D_1 and R_1 causing SCR_1 to conduct. When the positive half cycle is complete, the voltage will go to zero and SCR_1 will shut OFF. To switch the negative half cycle to the load, close S_2. See Figure 7-17. With S_2 closed, the gate will be connected for SCR_2 through diode D_2 and resistor R_2.

As the negative half cycle begins, SCR_2 becomes forward biased and gate current will flow through D_2 and R_1 causing SCR_2 to conduct. Once again, when the voltage crosses the zero voltage point, SCR_2 shuts OFF.

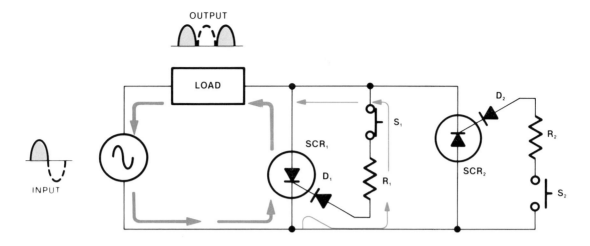

Figure 7-16. During the positive half of each cycle, SCR_1 is forward biased. If switch S_1 is closed, the gate current will flow through diode D_1 and resistor R_1, causing SCR_1 to trigger into conduction. At the end of each half cycle, the voltage will go to zero and SCR_1 will shut OFF.

hand is forward biased, and it allows capacitor C to charge to the polarity shown. No current flows through R since D_2 is in parallel with R. The forward-biased resistance of D_2 is so much lower than R that almost all current will pass through D_2 instead of through R. Since no current is flowing through the load, the oscilloscope will not indicate a voltage drop present.

NOTE: The oscilloscope on the negative half cycle would show only a slight movement of the trace. This is because the SCR is reverse biased and the current drawn through D_2, when charging C, is small.

When the positive alternation of the voltage source is applied, the circuit operation changes in the manner

as shown in Figure 7-19. When the positive alternation is applied to the circuit, the SCR will become forward biased such that it conducts forward current if a gate current of sufficient strength is present. The important factor is the amount of gate current. If the gate current is too small, the SCR will not fire. The amount of gate current in this circuit is controlled by variable resistor R and capacitor C. R and C form an RC network that determines the time of charge and discharge of the capacitor.

For C to charge on the positive alternation, it must first discharge the polarities it received on the negative alternation. Once discharged, it may recharge with the

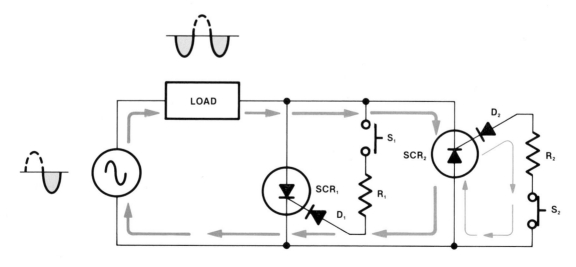

Figure 7-17. During the negative half of each cycle, SCR_2 is forward biased. If switch S_2 is closed, the gate current will flow through diode D_2 and resistor R_2, causing SCR_2 to conduct. At the end of each half cycle, the voltage goes to zero and SCR_2 shuts OFF.

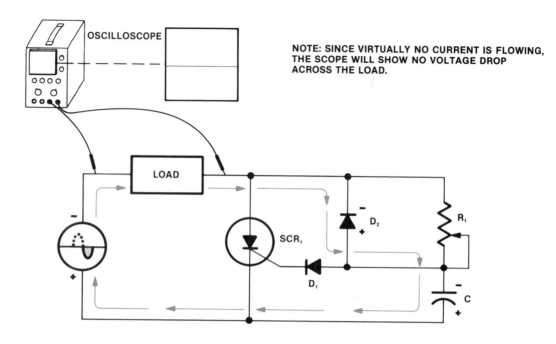

NOTE: SINCE VIRTUALLY NO CURRENT IS FLOWING, THE SCOPE WILL SHOW NO VOLTAGE DROP ACROSS THE LOAD.

Figure 7-18. On the negative half cycle, SCR_1 is reverse biased and does not conduct. Diode D_2 is forward biased, allowing capacitor C to charge to the polarity shown. No current is shown through resistor R_1 since D_2 is in parallel with R_1. D_2 is forward biased with such a low resistance that virtually all the current will go through D_2 rather than R_1.

opposite polarity. The charging and discharging rates are determined by R. The discharge and charge time of the RC network determines the time at which sufficient current is allowed to flow through the gate circuit to fire the SCR. If variable resistor R is in position A, or has zero resistance, the capacitor will discharge its reverse polarity almost immediately. Current would then be allowed to pass through D_1 which in turn fires the SCR. The result is that the entire positive alternation is allowed to pass through the SCR to the load as shown by the oscilloscope display pattern correlated to position A. If variable resistor R is moved to position B, where more resistance is placed in the RC network, the time of discharge for C will increase. The result is a delay in time for the gate current to reach its necessary value. Also, part of the positive alternation will be blocked as shown by the oscilloscope display pattern correlated to position B. As more resistance is added in positions C and D, the delay time is increased. The result is that the positive alternation will increasingly be blocked while a decreasing amount of current will be delivered to the load. Thus, the SCR then can deliver a varied output to the load on the positive alternation.

NOTE: To control the output in both the positive and negative alternations, two SCRs must be used.

SCR APPLICATIONS

In addition to being used in light dimmers and speed controls, the SCR is used in industry to control power in battery charges, power supplies, and machine tools. The SCR is used as a power control element in welders, power regulators, and temperature control systems.

Precision Heat Control

The SCR can be used in circuits to provide heat control. For example, it can bring a chemical mixture stored in a vat up to a specific temperature, and maintain that temperature. See Figure 7-20. With the proper circuitry, the temperature of the mixture can be precisely controlled. Using a bridge circuit, the temperature can be maintained within 1 °F over a temperature range of 20 °F to 150 °F. See Figure 7-21.

In the circuit of Figure 7-21, transformer *T1* has two secondary windings, *W1* and *W2*. *W1* furnishes voltage through the SCR to relay coil *K1*. *W2* furnishes AC voltage to the gate circuit of the SCR. Primary control over this circuit is accomplished through the use of the bridge circuit. The bridge circuit is formed by thermistor R_1, fixed resistors R_2 and R_3, and potentiometer R_4. Resistor R_5 is a current-limiting resistor used to protect the bridge circuit. The fuse is used to protect the primary of the transformer.

When the resistance of R_1 equals the resistance setting on R_4, the bridge is balanced. None of the AC voltage introduced into the bridge by winding *W2* is applied to the gate of the SCR. Hence, the relay coil *K1* remains de-energized and its normally closed contacts apply power to the heating elements.

If the temperature increases above a preset level, the resistance of thermistor R_1 decreases. The bridge becomes unbalanced such that a current flows to the gate of the SCR while the anode of the SCR is still positive. See Figure 7-22. This turns ON the SCR and

energizes the relay coil *K1*, thereby switching power from the load through the relay contact. If the temperature falls below the preset temperature setting, R_1 will unbalance the bridge in the opposite direction. Therefore, a negative signal is applied to the gate of the SCR when the anode of the SCR is positive. The negative signal stops the SCR from conducting and allows current to continue to flow to the heating elements.

NOTE: Locating the thermistor in the vat where the temperature must be controlled will provide the necessary feedback information.

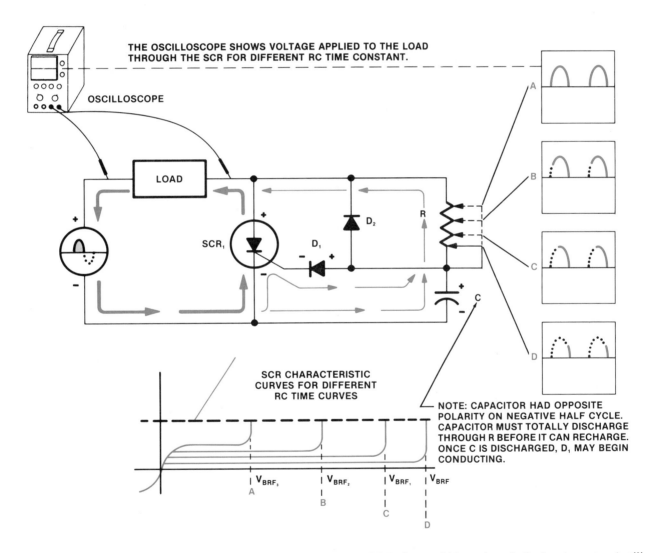

THE OSCILLOSCOPE SHOWS VOLTAGE APPLIED TO THE LOAD THROUGH THE SCR FOR DIFFERENT RC TIME CONSTANT.

OSCILLOSCOPE

LOAD

SCR₁   D₁   D₂   R

SCR CHARACTERISTIC CURVES FOR DIFFERENT RC TIME CURVES

NOTE: CAPACITOR HAD OPPOSITE POLARITY ON NEGATIVE HALF CYCLE. CAPACITOR MUST TOTALLY DISCHARGE THROUGH R BEFORE IT CAN RECHARGE. ONCE C IS DISCHARGED, D₁ MAY BEGIN CONDUCTING.

V_{BRF_3} A V_{BRF_2} B V_{BRF_1} C V_{BRF} D

Figure 7-19. On the positive half cycle of the AC input, the SCR is forward biased such that gate current will flow through diode D_1 and variable resistor R, causing the SCR to conduct. Since R and capacitor C form an *RC* time constant, the position of R determines the discharge time of C. If C is discharged rapidly due to position A, then D_1 will conduct earlier. The entire positive half cycle will then be delivered to the load. As resistance is added in positions B, C, and D, the time required for D_1 to conduct is increased. The result is decreased current delivered to the load.

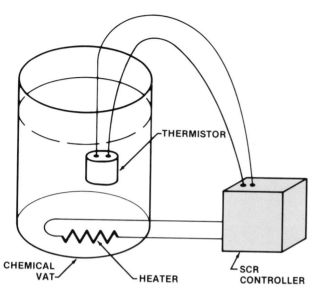

Figure 7-20. An SCR can be used in circuits to provide precision heat control. The SCR controller will bring the chemical mixture in a vat up to a specific temperature and maintain it within 1° over a temperature range from 20°F to 150°F.

Variable-Speed Fan Motor Controller

The SCR can be used in circuits to provide speed control to a fan motor. See Figure 7-23. By using a simple phase-control circuit, the SCR can provide a limited-range speed control. For a shaded-pole fan motor, the circuit can give a 3 to 1 speed control ratio. See Figure 7-24. In this circuit, diodes D_1, D_2, D_3, and D_4 form a full-wave bridge rectifier. This full-wave bridge rectifier converts the AC supply into a full-wave pulsating direct current (DC). The output of the full-wave is through the SCR, which exerts influence over both half cycles of the AC supply. See Figures 7-24 and 7-25. The oscilloscope in both figures shows the output across the motor for each cycle if the SCR is in full conduction.

Since an SCR can be triggered with as little as 0.5 milliamps of gate current, an inexpensive neon glow lamp can be used as the triggering device. Potentiometer R_1, timing capacitor C_3, and the neon glow lamp form the trigger circuit that determines the

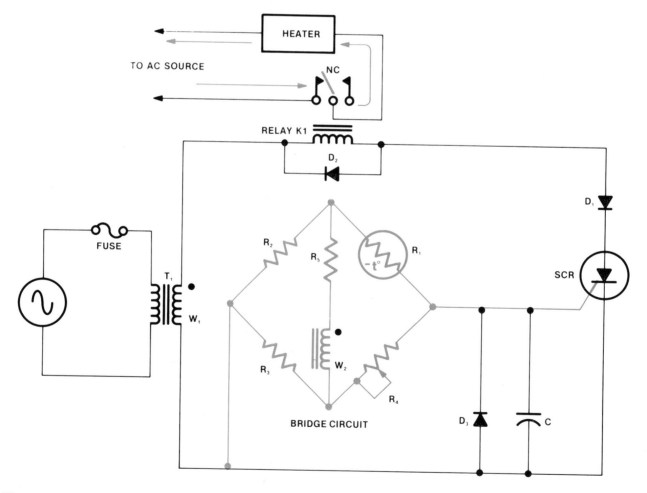

Figure 7-21. Voltage is provided to this circuit through transformer *T1*, which has two secondary windings, *W1* and *W2*. A bridge is formed by thermistor R_1, fixed resistors R_2, R_3, and potentiometer R_4. When the bridge is balanced, there is no gate current and relay *K1* remains de-energized, allowing the heater to stay ON.

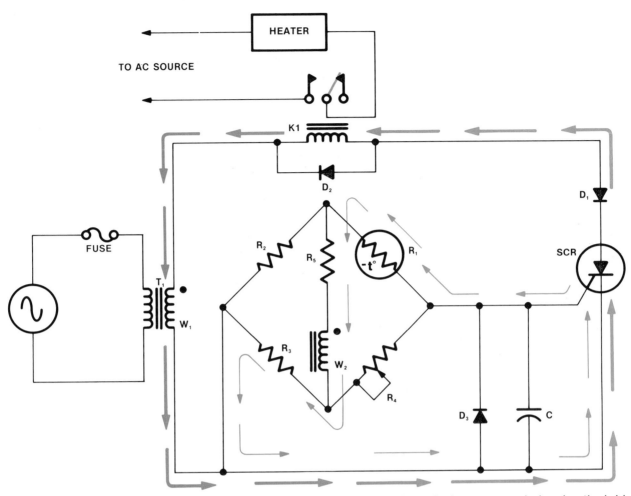

Figure 7-22. When the temperature increases, the resistance of thermistor R_1 decreases, unbalancing the bridge so that a gate current flows to the SCR while its anode is still positive. This turns ON the SCR and energizes the relay coil *K1*, thereby disconnecting power from the load.

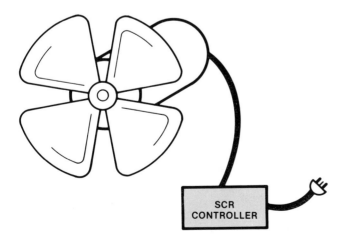

Figure 7-23. The SCR can be used in circuits to provide speed control for a fan motor.

SCR firing point in each half cycle. Reducing the resistance allows the SCR to fire early in each half cycle. Increasing the resistance retards the SCR, and the conduction in each half cycle is shortened. The longer the conduction is in each half cycle, the greater the motor RPM is. Diode D_5, connected across R_1, resets C_3 to zero volts each time the SCR triggers. C_1, C_2, R_2, and L_1 form a radio frequency interference filter that reduces noise and transients (high voltage variations) which, if present in larger amounts, would inadvertently trigger the SCR.

Time-Delayed Relay

The SCR can be used in circuits to provide time delay for motor control, photo enlargers, and other timed or delayed functions. In many situations, an SCR can be combined with a relay to form a *hybrid circuit*. The hybrid circuit turns a load ON and OFF after a preset

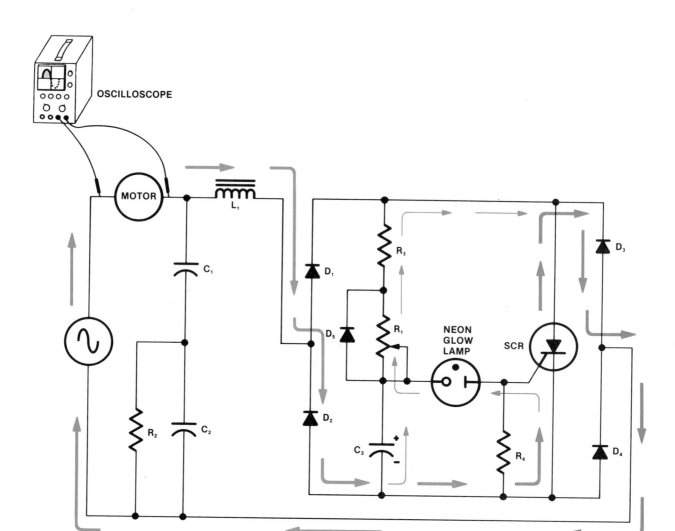

CURRENT FLOW DURING POSITIVE (+) HALF CYCLE

Figure 7-24. Depending upon which part of the conducting cycle the SCR fires, all, none, or part of the positive half cycle will be passed on to the load.

time delay. A hybrid circuit has an electronic component (SCR) and an electromechanical component (relay) in the same circuit.

A typical hybrid circuit is shown in Figure 7-26. The operation of a hybrid circuit starts with switch S_1 in the RESET position. Capacitor C quickly charges to the peak negative value of the supply voltage (approximately 170 volts) and effectively shuts OFF any gate current to the SCR. In this position, the load is OFF. S_1 turned to the TIME position, as shown in Figure 7-27, turns ON the load. Capacitor C starts to discharge toward zero at a rate determined by the setting of potentiometer R_2. Since a delay is built into this circuit, the RC time constant will be long. The delay may be shortened by reducing the resistance of R_2. When the voltage across C finally drops to zero,

reverses, and reaches about one volt, the SCR triggers relay coil $K1$ to turn OFF the load.

Car Alarm Circuit

The SCR can be used as part of a security alarm system for cars or homes. See Figure 7-28. By using diodes, switches, and an SCR firing circuit, visual and audible alarms can be activated to serve as warning signals in an attempted theft. By placing diodes in series with the ignition coil, radio, or tape player, each car accessory serves as a means for triggering the SCR when it is powered up. If normally open switches are installed as shown, the list can include the trunk, hood, or a pressure-sensitive switch under the floor mat.

The operation of the security alarm system circuit

in Figure 7-28 is simple. If both the arm switch S_1 and the reset switch S_2 are closed and the ignition switch is turned ON (either by a key or a jumper wire), current flows through diode D_1 (associated with the ignition switch) and R_1. This puts a charge on capacitor C. When the charge on C is sufficient to turn ON the transistor, the current through the transistor also turns ON the SCR. The SCR supplies current to a horn or siren. Once the SCR is ON, it remains ON regardless of the condition of the diodes or the transistor. Only the opening of the reset switch S_2, which is concealed within the vehicle, turns OFF the SCR. The values of R_1 and C in the timing network are selected to provide sufficient delay for the vehicle owner to enter the vehicle and open the reset switch and the arm switch. The arm switch is left open when the vehicle is in use. It is closed when the vehicle is unattended.

SCR MOUNTING AND COOLING

Proper operation of an SCR depends to a great extent on the correct mounting and cooling of the device. If the temperature of an SCR is allowed to rise too much, it may fail because of thermal runaway. Circuits may also change characteristics because of insufficient cooling, which reduces the forward breakover voltage. For these reasons, most SCRs are designed with some type of heat transfer mechanism to dissipate internal heat loss.

Mounting surfaces, like heat sinks, are generally an integral part of the heat transfer path of an SCR. Proper mounting is always essential for successful SCR cooling. Incorrect mounting and cooling of the SCR results in the same problem and therefore must be treated together.

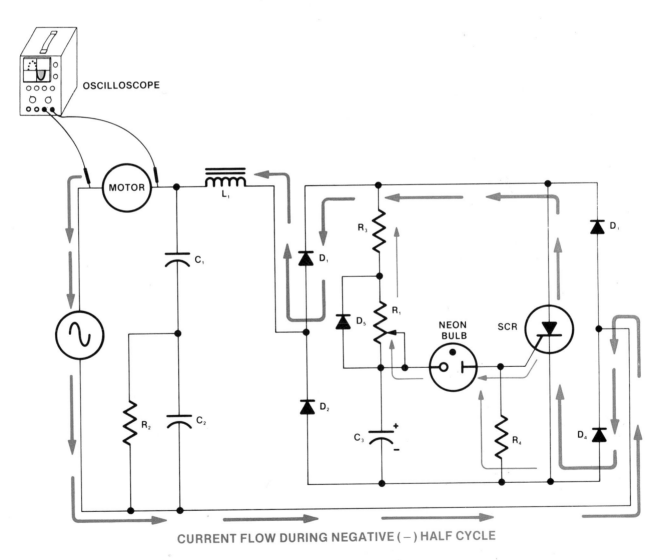

CURRENT FLOW DURING NEGATIVE (−) HALF CYCLE

Figure 7-25. Depending upon which part of the conducting cycle *SCR*₁ fires, all, none, or part of the negative half cycle will be passed on to the load.

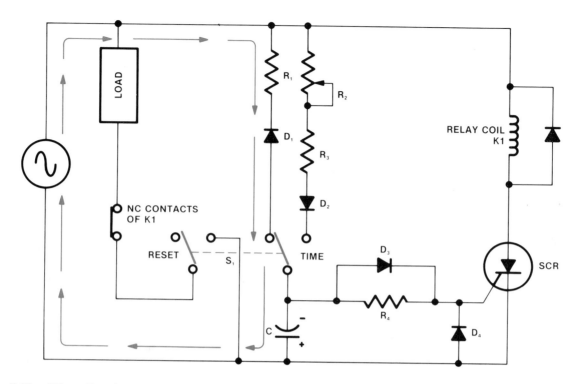

Figure 7-26. When the time-delay circuit is in the RESET position, capacitor *C* quickly charges to the peak negative value of the supply voltage, effectively shutting OFF any gate current to the SCR.

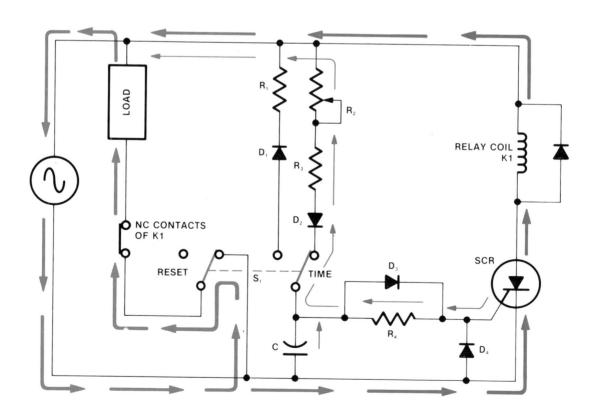

Figure 7-27. Turning switch S_1 to the TIME position turns ON the load. Capacitor *C* starts to discharge toward zero at a rate determined by the *RC* time constant. When the voltage across *C* finally drops to zero, reverses, and reaches about one volt, the SCR triggers the relay coil *K1* to turn OFF the load.

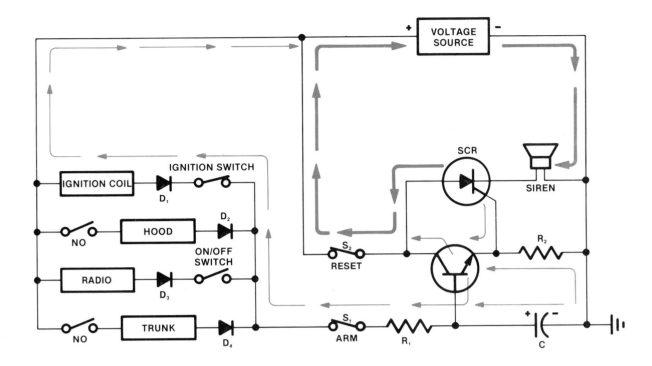

Figure 7-28. If both the arm switch S_1 and the reset switch S_2 are closed, and the ignition switch is turned ON, current will flow through diode D_1. (D_1 is associated with the ignition switch and R_1.) This puts a charge on capacitor C. When the charge on C is sufficient to turn ON the transistor, the current through the transistor also turns ON the SCR. Once the SCR is ON, only opening the reset switch will turn OFF the SCR.

Lead-Mounted SCRs

For small lead-mounted SCRs, cooling is maintained by radiation and convection from the surface of the SCR case. Cooling is also maintained by thermal conduction through the SCR leads. See Figure 7-29.

The following practices for minimizing the SCR temperature should be observed whenever possible:

1. Minimum lead length to the terminal board, socket, or PC board should be used because it permits the mounting points to assist in the cooling of the SCR.

2. Other heat-dissipating elements such as power resistors should not be connected directly to the SCR leads where avoidable.

3. High temperature devices such as lamps, power transformers, or resistors, should be shielded from radiating their heat directly on the SCR case.

4. In the final mounting, be sure insulators are in position and that silicon paste is applied to both sides for maximum heat transfer.

Press-Fit SCRs

Many medium-current SCRs utilize the press-fit package. See Figure 7-30. This package is designed primarily for forced insertion into a slightly undersized

hole in the heat exchanger (heat sink). When properly mounted, the press-fit SCR has a lower thermal drop to the heat exchanger than the stud-mounted SCR. Also, in high volume applications, the cost of the press-fit SCR is generally less than the cost of the stud-mounted SCR.

The procedure for mounting a press-fit SCR is:

1. The hole may be punched and reamed in a flat plate, extruded and sized in sheet metal, or drilled and reamed. A slight chamfer on the hole should be used to guide the housing.

2. To ensure maximum heat transfer, the entire knurl should be in contact with the heat exchanger. The SCR must not be inserted into the heat exchanger past its knurl. This is to prevent the heat sink from taking pressure off the knurl in a deep hole.

3. The insertion force must be limited as specified by the manufacturer. Following the manufacturer's specification helps prevent misalignment with the hole. Excessive SCR-to-hole interference will also be avoided. Pressure must be uniformly applied to the face of the header when inserting the SCR.

NOTE: Heat exchanger materials may be (in order of preference) copper, aluminum, or steel. The heat exchanger thickness should be a minimum of 1/8" the width of the knurl on the housing.

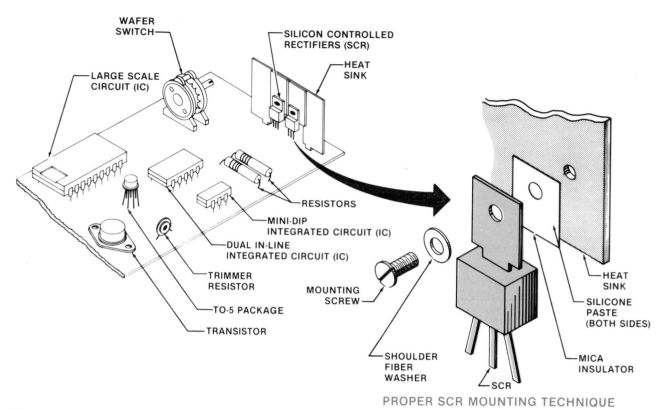

Figure 7-29. For small lead-mounted SCRs, cooling is maintained by radiation and convection from the surface of the SCR case and by thermal conduction through the SCR leads.

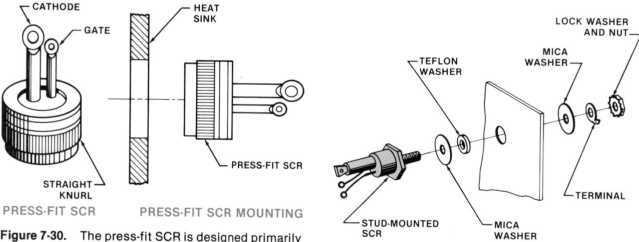

Figure 7-30. The press-fit SCR is designed primarily for forced insertion into a slightly undersized hole in the heat exchanger (heat sink).

Figure 7-31. SCRs, like regular silicon diodes, can be stud-mounted.

Stud-Mounted SCRs

When installing stud-mounted SCRs, the same procedures should be followed as for mounting stud-mounted diodes. See Figure 7-31. (Refer to Chapter 3, SEMICONDUCTOR DIODES, under DIODE INSTALLATION AND SERVICE.)

Flat-Pack and Press-Pack SCRs

The advantage of flat-pack and press-pack SCRs is the increased surface area available for thermal conduction. See Figure 7-32. At a given temperature, their large, smooth flat surface allows greater power dissipation than the conventional stud-mounted SCRs. The major disadvantage is the cost of the more elaborate mounting procedure.

A press-pack SCR must be properly installed to provide maximum heat transfer. See Figure 7-33. Uneven force distribution results in poor thermal and electrical contact, which will seriously impair SCR performance.

CAUTION: Information provided by manufacturers of SCRs on installation must be followed closely.

TESTING AN SCR

Special test equipment, such as an oscilloscope, is needed to properly test an SCR under operating conditions. However, a rough test, using a test circuit, can be made with an ohmmeter. See Figure 7-34. The procedure for a rough test is:

1. Place the ohmmeter on the ''R × 100'' scale.
2. Connect the negative lead of the ohmmeter to the cathode.
3. Connect the positive lead of the ohmmeter to the

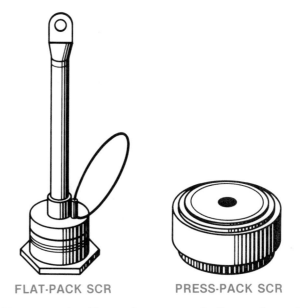

FLAT-PACK SCR   PRESS-PACK SCR

Figure 7-32. The advantage of flat-pack and press-pack SCRs is the increased surface area available for thermal conduction.

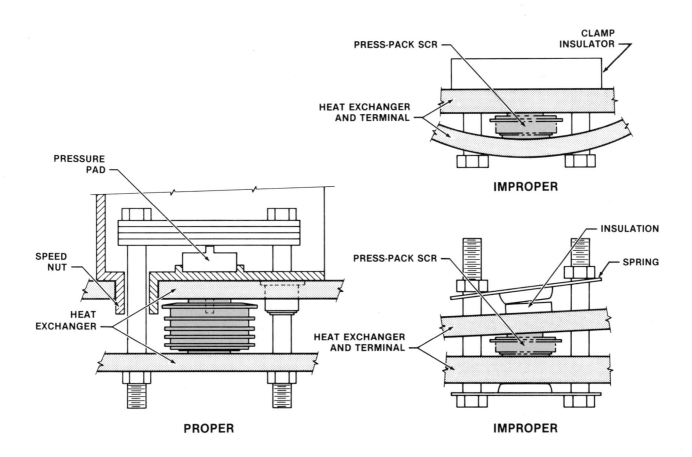

Figure 7-33. A press-pack SCR must be properly installed to provide maximum heat transfer.

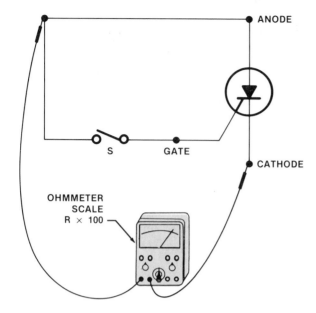

Figure 7-34. An SCR can be tested, using a test circuit, with an ohmmeter. Using the connections shown, the ohmmeter should show a high resistance when switch S is open and a low resistance when S is closed.

anode. The resistance reading should be so high that the ohmmeter indicates infinity. (The resistance will in fact be over 250,000 ohms.)

4. Short-circuit the gate to the anode by closing switch S.

5. The resistance reading should be so low that the ohmmeter will indicate almost zero. (The resistance will be about 10 to 50 ohms, but this will not register on the "R × 100" scale.)

6. When switch S is opened, the zero resistance reading should remain.

7. Remove and reconnect the ohmmeter leads with the positive lead to the cathode and the negative lead to the anode. The resistance reading should be so high that the ohmmeter indicates infinity. (The resistance will in fact be over 250,000 ohms.)

8. Short-circuit the gate to the anode by closing switch S. The resistance should remain high since the SCR is reverse-biased and cannot conduct.

9. Open switch S. The high resistance should remain since the SCR is reverse-biased and has no gate current.

NOTE: If the SCR does not respond as indicated for each of these steps, it is defective and must be replaced.

Chapter 7 - Review Questions

1. How is the physical size of an SCR affected by an increase in current rating?
2. Which electrode on an SCR is the control point for the SCR?
3. What effect does gate current have on the level of breakover voltage?
4. Name the three electrodes of an SCR.
5. What is forward blocking current?
6. How many layers of P-type and N-type material does an SCR have?
7. What are the two operating states of an SCR?
8. What two events can cause an SCR to conduct in the forward direction?
9. What devices could be connected together to form an SCR equivalent circuit?
10. What is the SCR most often used for?
11. What is a hybrid circuit?
12. Which type of component can be used to reverse bias an SCR to turn it OFF?
13. Define phase control.
14. What type of circuit can be used to provide a time delay in an SCR circuit?
15. In what types of circuits are SCRs found?
16. What happens to an SCR if it overheats?
17. What precautions should be taken when lead mounting SCRs?
18. What are the advantages of press-fit SCRs?
19. What is the advantage of flat-pack and press-pack SCRs?
20. What is the procedure for testing an SCR?

8 TRIAC, DIAC, AND UNIJUNCTION TRANSISTOR

Triacs, diacs, and unijunction transistors (UJTs), along with silicon controlled rectifiers (SCRs), are often found in the same circuitry. Triacs and SCRs are control devices. Diacs and UJTs form the triggering circuits for triacs and SCRs.

Key Words

Base 1	Pedestal voltage	Terminal 2 (T_2)
Base 2	Reed switch	Triac
Bidirectional	Saturation	Unijunction transistor (UJT)
Diac	Terminal 1 (T_1)	Valley current point
Peak current point		

TRIAC

A *triac* is a three-electrode AC semiconductor switch. It is triggered into conduction in both directions by a gate signal in a manner similar to the action of an SCR. The triac was developed to provide a means for producing improved controls for AC power. Triacs are available in a variety of packaging arrangements. See Figure 8-1. They can handle a wide range of amperages and voltages. Triacs generally have relatively low current capabilities compared to SCRs. Triacs are usually limited to less than 50 amps and cannot replace SCRs in high-current applications.

Triac Construction

Although triacs and SCRs look alike, their schematic symbols are different. See Figure 8-2. The terminals of the triac are the gate, *Terminal 1 (T_1)*, and *Terminal 2 (T_2)*. There is no designation of anode and cathode.

Current may flow in either direction through the main switch terminals, T_2 and T_1. Terminal T_2 is the reference terminal for all voltages. T_2 is the case or metal-mounting tab to which the heat sink can be attached. The structure of the triac is complex. However, for all practical purposes, the triac can be considered two NPN switches sandwiched together on a single N-type material wafer. See Figure 8-3.

Triac Operation

The triac blocks current in either direction between T_1 and T_2. A triac can be triggered into conduction in either direction, by a momentary pulse in either direction, supplied to the gate. The triac operates much like a pair of SCRs connected in a reverse parallel arrangement. See Figure 8-4. If the appropriate signal is applied to the triac gate, it will conduct.

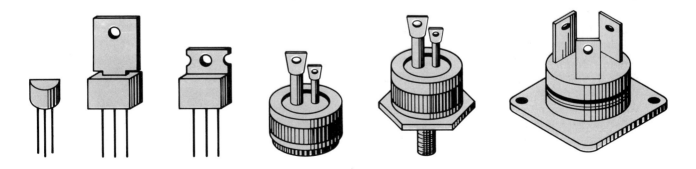

Figure 8-1. Triacs are available in a variety of packaging arrangements. They can handle a wide range of amperages and voltages.

119

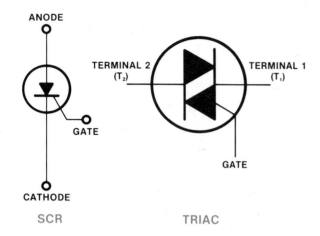

Figure 8-2. The triac symbol has no designation of anode and cathode like the SCR symbol has. The terminals of a triac are the gate, Terminal 1 (T_1), and Terminal 2 (T_2).

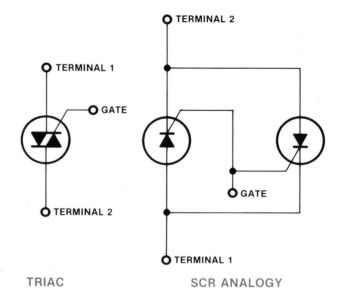

Figure 8-4. The triac operates much like a pair of SCRs connected in a reverse parallel arrangement.

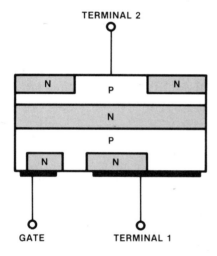

Figure 8-3. The triac can be considered two NPN switches sandwiched together on a single N-type material wafer.

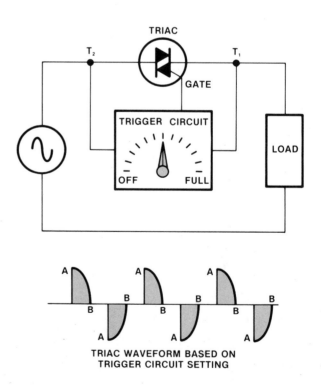

Figure 8-5. The triac remains OFF until the gate is triggered at point A. At point A, the trigger circuit pulses the gate and turns ON the triac, allowing current to flow. At point B, the forward current is reduced to zero and the triac turns OFF.

Figure 8-5 shows how a triac may be triggered into conduction. The triac remains OFF until the gate is triggered at point A. At point A, the trigger circuit pulses the gate and turns ON the triac, allowing current to flow. At point B, the forward current is reduced to zero, so the triac turns OFF. The trigger circuit can be designed to produce a pulse that varies at any point in the positive or negative half cycle. Therefore, the average current supplied to the load can vary.

One advantage of the triac is that virtually no power is wasted by being converted to heat. Heat is generated when current is impeded, not when current is switched OFF. The triac is either fully ON or fully OFF. It never

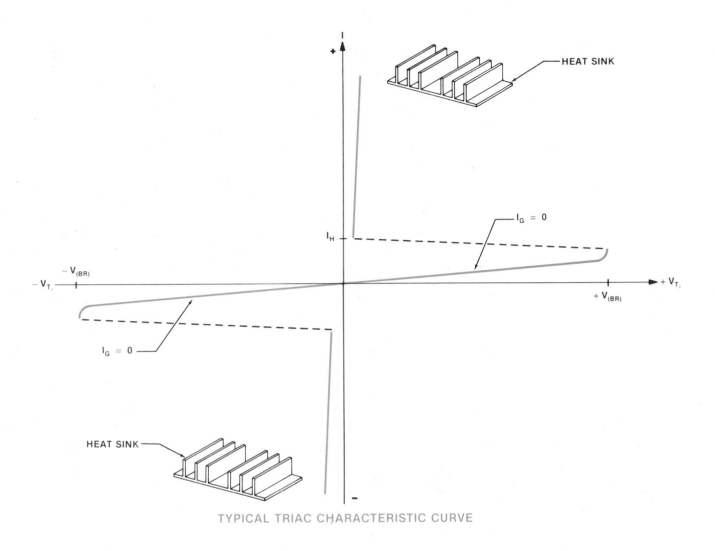

TYPICAL TRIAC CHARACTERISTIC CURVE

Figure 8-6. The characteristics of the triac are based on Terminal 1 as the voltage reference point. The triac can be triggered into conduction in either direction by a gate current (I_G) in either direction.

partially limits current. Another important feature of the triac is the absence of a reverse breakdown condition of high voltages and high current, such as those found in diodes and SCRs. If the voltage across the triac goes too high, the triac will merely turn ON. When turned ON, the triac can conduct a reasonably high current.

Triac Characteristic Curve

The characteristics of the triac are based on Terminal 1 (T_1) as the voltage reference point. See Figure 8-6. The polarities shown for voltage and current are the polarities of Terminal 2 (T_2) with respect to T_1. The polarities shown for the gate are also with respect to T_1. Again, the triac may be triggered into conduction in either direction by a gate current I_G in either direction.

Triac Applications

Because of their versatility, triacs are often used instead of mechanical switches. Also, where amperage is low, triacs are more economical than back-to-back SCRs.

Single-Phase Motor Starter. Often, a capacitor-start or split-phase motor must operate where arcing of a mechanical cut-out start switch is undesirable or even dangerous. They must operate, for example, in a place where explosive fumes are present near the motor. In such cases, the mechanical cut-out start switch may be replaced with a triac. See Figure 8-7.

The gating and cut-out signal is given to the triac through the current transformer. As the motor speeds up, the current reduces in the current transformer and no longer triggers the triac. With the triac OFF, the starter windings are removed from the circuit.

Another method of sending signals to the triac is by using a tachometer generator. A tachometer generator puts out a specific voltage for a certain amount of RPM. As RPM increases, voltage increases. As RPM decreases, voltage decreases. The signal would be sent to a trigger circuit which would fire the triac at a preset RPM. With this arrangement, the drop-out or cut-out could be precisely set.

Triac Replacement of Reversing Contactors. Another use for triacs is as a replacement for reversing contactors. When used with a reversing permanent split capacitor motor, a pair of triacs can provide a rapidly responding, reversing motor control circuit. See Figure 8-8. S_1 and S_2 can be *reed switches*. Reed switches consist of one movable and one fixed arm. As a magnetic field is placed next to the reed switch, the movable arm (reed) is positioned to open or close the circuit. Triacs can also be gated by mechanical switches or other solid state switches.

In this particular circuit, it is important to ensure that the triacs have a sufficient voltage rating. If one triac is ON, the other triac must block the voltage. This voltage is created by line voltage due to the inductance and capacitance between the motor winding and the capacitor.

NOTE: As a rule of thumb, the triac voltage rating should be 1.5 times the voltage rating of the capacitor.

Automatic Night Light. The triac can be used in a home or business security system circuit as a means of turning lights ON at night and OFF at dawn. See Figure 8-9. The voltage-divider circuit divides the DC voltage between R_2 and the parallel combination of the photocell, with R_3 in series with coil L_1. When the photocell is exposed to light, its resistance drops to a few ohms, thereby depriving L_1 of enough current to close the reed switch S. With S open, the triac will not conduct and the light will remain OFF. If no light falls on the photocell, its resistance increases to several megaohms. The current from R_2 then passes through

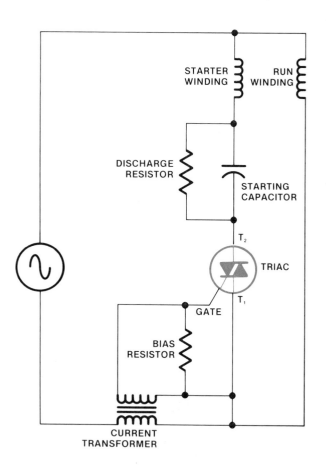

Figure 8-7. A mechanical cut-out switch may be replaced with a triac. As the motor speeds up, the current drops off in the current transformer and the transformer no longer triggers the triac. With the triac OFF, the starter windings are removed from the circuit.

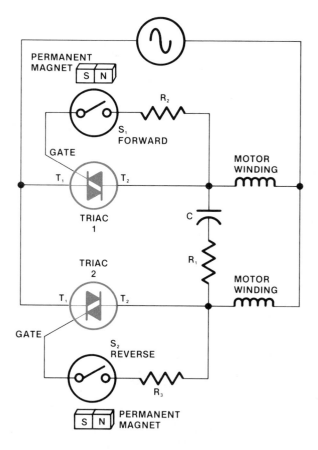

Figure 8-8. A pair of triacs can replace reversing contactors on a permanent split capacitor motor. The triacs are gated in this circuit by reed switches S_1 and S_2.

R_3 and coil L_1. This current causes the reed switch S to close, and energizes the triac by passing current to the gate. Once the triac fires, the light turns ON.

Testing a Triac

Special test equipment, such as an oscilloscope, is needed to test a triac under operating conditions. However, a rough test can be made, using a test circuit, with an ohmmeter. See Figure 8-10. To make a rough test:

1. Place the ohmmeter on the "R × 100" scale.
2. Connect the negative lead of the ohmmeter to Terminal 1.
3. Connect the positive lead of the ohmmeter to Terminal 2. The resistance will be so high that the ohmmeter indicates infinity. The resistance will, in fact, be over 250,000 ohms.

4. Short circuit the gate by closing switch S. The resistance reading should be so low that the ohmmeter will indicate almost zero. (The resistance will be about 10–50 ohms, but will not register on the "R × 100" scale.)

5. When the gate switch is opened, the zero resistance reading should remain.

6. Remove and reconnect the ohmmeter leads with the positive lead to Terminal 1 and the negative lead to Terminal 2. The resistance will be so high that the ohmmeter indicates infinity. The resistance will, in fact, be over 250,000 ohms.

7. Short circuit the gate by closing switch S. The resistance reading should be so low that the ohmmeter will indicate almost zero. (The resistance will be about 10–50 ohms, but will not register on the "R × 100" scale.)

8. When the gate switch is opened, the zero resistance reading should remain.

NOTE: If the triac does not respond as indicated in each of these steps, it is probably defective and must be replaced.

DIAC

Figure 8-11 shows a typical packaging of a *diac* and its schematic symbol. The diac is a three-layer, *bidirectional* device. See Figure 8-12. Unlike the transistor, the two junctions are heavily and equally doped. Each junction is almost identical to the other.

Diac Operation

Electrically, the diac acts much like two zener diodes that are series connected in opposite directions. See

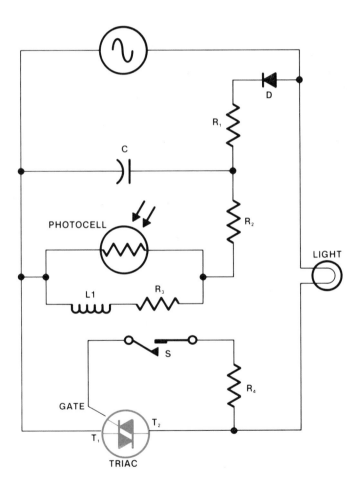

Figure 8-9. The automatic night light is triggered by the photocell and reed switch S. When no light falls on the photocell, its resistance is high and most of the current flows through coil L_1. With current through L_1, the triac fires and causes the light to glow. When light is shining on the photocell, its resistance decreases, causing most of the current to pass through the photocell. The reduction in the current through L_1 opens the reed switch contacts and the triac turns OFF the light.

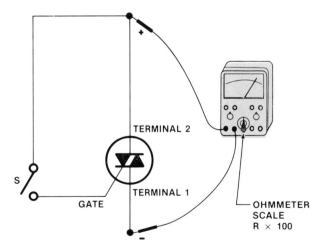

Figure 8-10. An ohmmeter can be used to make a rough test of the operating condition of a triac.

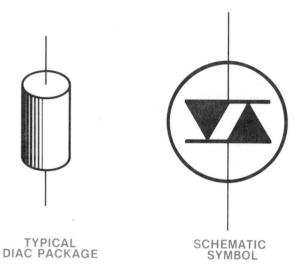

TYPICAL
DIAC PACKAGE   SCHEMATIC
                            SYMBOL

Figure 8-11. Shown above is a typical diac package and its schematic symbol.

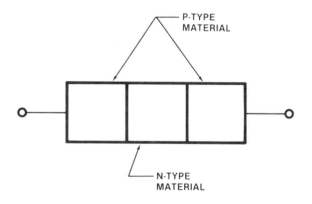

Figure 8-12. The diac is a three-layer, bidirectional device. Unlike the transistor, the two junctions are heavily and equally doped.

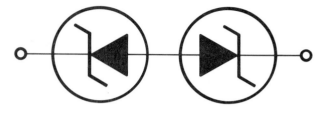

DIAC EQUIVALENT

Figure 8-13. Electrically, the diac acts much like two zener diodes that are series connected in opposite directions.

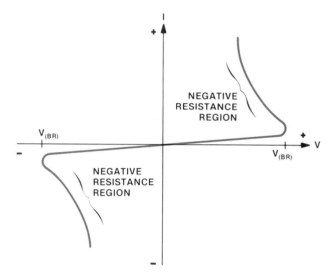

Figure 8-14. The diac is bidirectional and is used primarily as a triggering device. When a positive or negative voltage reaches the breakover voltage, the diac rapidly swtiches from a high-resistance state to a low-resistance state.

Figure 8-13. The diac is used primarily as a triggering device. It accomplishes this through the use of its negative resistance characteristic. (Negative resistance characteristic means the current decreases with an increase of applied voltage.) The diac has negative resistance because it does not conduct current until the voltage across it reaches breakover voltage. See Figure 8-14. When a positive or negative voltage reaches the breakover voltage, the diac rapidly switches from a high-resistance state to a low-resistance state. Since the diac is a bidirectional device, it is ideal for controlling triacs, which are also bidirectional.

Diac Applications

The gate-control circuits of triacs can be improved by adding a breakover device in the gate lead. The diac is a breakover device. Using a diac in the gate-triggering circuit offers an important advantage over simple *RC* gate-control circuits. The advantage is that the diac delivers a pulse of gate current rather than a sinusoidal gate current. This results in a better controlled firing sequence. Thus, diacs are used almost exclusively as triggering devices.

Triac/Diac Light Dimmer. A diac can be used to trigger a triac in a light control circuit. See Figure 8-15. When the variable resistor R_2 is at its lowest value, capacitor C charges rapidly at the beginning of each half cycle of the AC voltage. When the voltage across C reaches the breakover voltage of the diac (about 30 volts), C is discharged into the gate of the triac. The triac fires and conducts heavily until the voltage returns to zero. Thus, the triac is ON early in each half cycle. The triac continues to conduct from the time it is triggered to the end of each half cycle. Therefore, current will flow through the lamp for most of each half cycle and produce a full brightness.

As the resistance of R_2 increases, the time required to charge C to the breakover voltage of the diac increases. This causes the triac to fire later in each half cycle. Therefore, the length of time that current flows through the lamp is reduced.

Triac/Diac Universal Motor-Speed Controller. A triac/diac combination can be used to control the power to a universal motor. See Figure 8-16. In this circuit, capacitor C charges up to the firing voltage of the diac in either direction. Once fired, the diac applies a voltage to the gate of the triac. The triac conducts and supplies power to the motor.

NOTE: The triac will conduct in either direction.

Since the universal motor is basically a series DC motor, current flowing in either direction will cause rotation in only one direction. Speed may be changed by varying the resistance of potentiometer R, which in turn, varies the RC time constant.

Testing a Diac

The ohmmeter can be used to test a diac for an open circuit. See Figure 8-17. The procedure used to test for an open circuit is:

1. Place the ohmmeter on the "R × 100" scale.

2. Connect the ohmmeter leads as shown in A of Figure 8-17, and record the resistance reading.

3. Reconnect the ohmmeter leads as shown in B of Figure 8-17, and record the resistance reading.

NOTE: Since the diac is essentially two diodes connected in parallel, both readings should show low resistance. Also, testing a diac in this manner will only show that the component is open. If it is suspected that the diac is open, a second test, using an oscilloscope, should be performed.

The procedure for testing for an open diac, using an oscilloscope is (See Figure 8-18):

1. Set up the test circuit.

2. Apply power to the circuit.

3. Adjust the oscilloscope.

NOTE: If the diac is good, a trace similar to that shown in Figure 8-18 will appear.

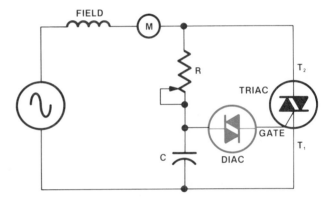

Figure 8-16. In this circuit, capacitor C charges up to the firing voltage of the diac in either direction. Once fired, the diac will apply voltage to the gate of the triac. The triac will conduct and apply power to the motor. Speed may be changed by varying the resistance of potentiometer R, which in turn, varies the RC time constant.

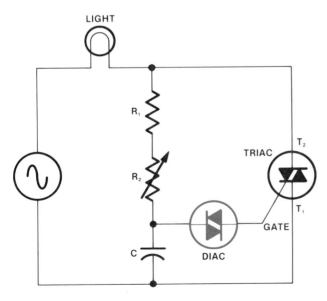

Figure 8-15. The triac in this circuit is triggered by the diac to provide control over the light intensity. The diac, in turn, is controlled by the RC time constant that is developed by variable resistor R_2 and capacitor C. The lower the value of R_2 is, the less time the triac stays ON. Therefore, the brightness of the light is reduced.

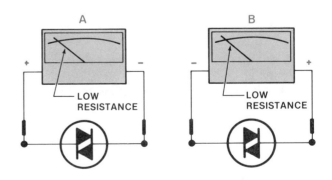

Figure 8-17. An ohmmeter can be used to test a diac for an open circuit. If the diac is good, the resistance reading will be low in either direction.

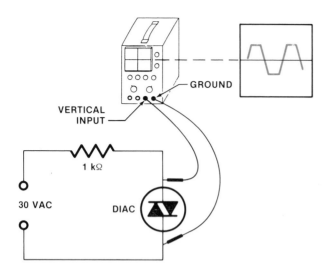

Figure 8-18. An oscilloscope and test circuit can also be used to test a diac. If the diac is operating properly, the oscilloscope will display the waveform shown.

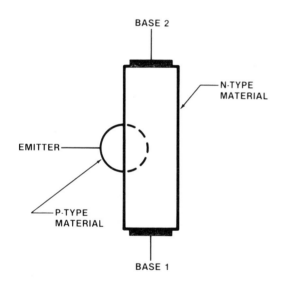

Figure 8-19. The unijunction transistor (UJT) consists of a bar of N-type material with a region of P-type material doped within the N-type material. The lead extending from the P-type material is the emitter. The other two leads are Base 1 (B_1) and Base 2 (B_2), respectively.

UNIJUNCTION TRANSISTOR (UJT)

The *unijunction transistor (UJT)* consists of a bar of N-type material with a region of P-type material doped within the N-type material. See Figure 8-19. The N-type material functions as the base and has two leads, Base 1 (B_1) and Base 2 (B_2). The lead extending from the P-type material is the emitter (E). The schematic symbol for a UJT is shown in Figure 8-20, left.

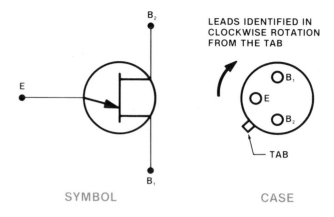

Figure 8-20. In the schematic symbol for a UJT, the arrowhead represents the emitter. It always points to B_1.

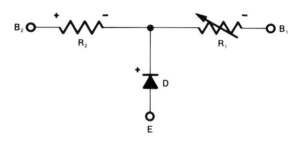

Figure 8-21. The equivalent circuit of a UJT can be thought of as a diode connected to a voltage divider network.

In the schematic symbol, the arrowhead represents the emitter (E). Although the leads are usually not labeled, they can be easily identified because the arrowhead always points to B_1. The case layout and lead identification for the UJT are shown in Figure 8-20, right.

The UJT is used primarily as a triggering device because it serves as a step-up device between low-level signals and SCRs and triacs. Outputs from photocells, thermistors, and other transducers can be used to trigger UJTs, which in turn, fire SCRs and triacs. UJTs are also used in oscillators, timers, and voltage-current sensing applications.

UJT Equivalent Circuit

It is helpful to study the equivalent circuit of a UJT in order to understand how a UJT operates. See Figure 8-21. The UJT can be considered a diode connected to a voltage divider network. Variable resistor R_1 and fixed resistor R_2 form the voltage divider network in the UJT. The PN junction creates the diode, and the N-type material serves as the resistive voltage divider.

UJT Biasing

In normal operation, B_1 is grounded, and a positive voltage is applied to B_2. See Figure 8-22. The internal resistance between B_1 and B_2 will divide at the emitter (*E*), with approximately 60% of the resistance between *E* and B_1. The remaining 40% of resistance is between *E* and B_2. The net result is an internal voltage split. This split provides a positive voltage at the N-type material of the emitter junction, creating an emitter junction that is reverse biased. As long as the emitter voltage remains less than the internal voltage, the emitter junction will remain reverse biased even at a very high voltage.

However, if the emitter voltage rises above this internal value, a dramatic change takes place. When the emitter voltage is greater than the internal value, the junction becomes forward biased. Also, the resistance between *E* and B_1 drops rapidly to a very low value. The UJT characteristic curve shows the dramatic change in voltage due to this resistance change. See Figure 8-23.

The action of the emitter-B_1 resistance is much like that of the variable resistor R_1 of the equivalent circuit.

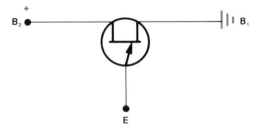

Figure 8-22. In normal operation, B_1 is grounded and a positive voltage is applied to B_2.

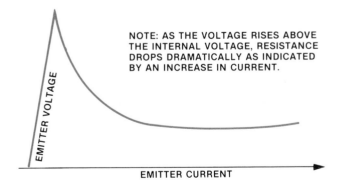

NOTE: AS THE VOLTAGE RISES ABOVE THE INTERNAL VOLTAGE, RESISTANCE DROPS DRAMATICALLY AS INDICATED BY AN INCREASE IN CURRENT.

Figure 8-23. When the emitter voltage rises above the internal voltage, the junction becomes forward biased and the resistance between the emitter and B_1 drops rapidly to a very low value. This characteristic curve shows the dramatic change in voltage due to the resistance change.

See Figure 8-24. R_1 is normally higher than R_2, but collapses to a very low value when the emitter is forward biased.

Theory of Operation

As long as the emitter-B_1 junction is reverse biased and no current flows into the emitter, the current flow in the N-type material will be minimal because of the small amount of doping that creates a high resistance. When the emitter-B_1 junction is forward biased, the junction turns ON, causing carriers to be injected into the base region. These carriers create an excess of holes. Their presence in the N-type material increases conductivity. Consequently, the resistance of the region is lowered. Once started, current flows easily between B_1 and *E*. Thus, the conductivity of this region is controlled by the flow of emitter current.

Saturation

Once the emitter-B_1 junction is forward biased, the lowered resistance across the emitter-B_1 junction means less voltage drop will appear at this junction. With this lower voltage drop, the junction has an increased forward bias voltage. The increased forward bias voltage lowers the junction resistance further. Eventually, the emitter circuit saturates and no additional increase in current results. If the voltage at *E* is increased still further, the UJT moves into its saturation region and the current increases only gradually. See Figure 8-25. The region between the *peak current point* and *valley current point (I)* is the negative-resistance region. In the negative-resistance region, the UJT has its greatest application as a triggering device.

Triggering Circuit

A typical use of a UJT is as a triggering circuit for a triac or similar device. Figures 8-26 and 8-27 show the basic operation of a UJT triggering circuit. When

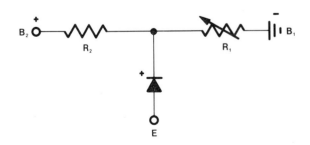

Figure 8-24. The action of the emitter-B_1 resistance is much like that of variable resistor R_1. R_1 is normally higher than R_2, but collapses to a very low value when the emitter is in the forward bias condition.

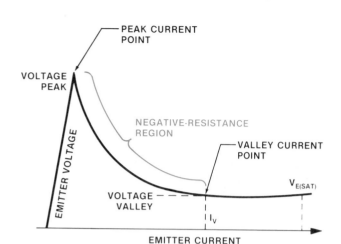

Figure 8-25. The region between the peak current point and the valley current point is the negative-resistance region. In the negative-resistance region, the UJT has its greatest application as a triggering device.

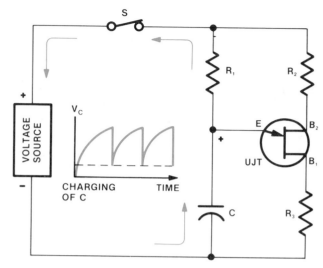

Figure 8-26. As capacitor *C* begins to charge through resistor R_1, the voltage across *C* will begin to increase.

switch *S* is closed, the voltage-divider action of the UJT produces a voltage between B_1 and the N-type material of the emitter junction. At this same instant, the emitter voltage will be zero since it is tied to capacitor *C*. The emitter junction at that point is reverse biased, and no current flows through the junction. As *C* begins to charge through resistor R_1, the voltage across *C* will begin to increase. See Figure 8-26.

To forward bias the emitter, the emitter must be more positive than the base (+.6 volt for silicon, +.2 volt for germanium). Assuming a silicon crystal is used in the UJT, the junction will become forward biased when the control voltage reaches 0.6 volts beyond the junction voltage. With the junction forward biased, the internal resistance of the emitter-B_1 region will drop dramatically. This causes *C* to discharge its energy through base load resistor R_3. See Figure 8-27, left. Once the capacitor has discharged enough to reduce the forward bias on the junction, the resistance of the junction will return to normal. The cycle (the capacitor charging and discharging) will then repeat itself.

Each time the emitter becomes forward biased, the total resistance between B_1 and B_2 drops, permitting an increase in current through the UJT. As a result, a positive pulse (V_B) will appear at B_1, and a negative pulse (V_B) will appear at B_2 at the time the capacitor discharges. See Figure 8-27, right.

NOTE: The repetition rate, or frequency, of the discharge voltage is determined by the values of R_3 and C. Increasing either one makes the device run more slowly. The pulses that appear across B_1 and B_2 are very useful in triggering SCRs and triacs.

UJT Applications

The UJT is not used as a transistor amplifier, but it is used in switching and timing applications. The use of the UJT often reduces the number of components needed to perform a given function. The component reduction is often less than half that required when using bipolar transistors. Since UJTs are relatively inexpensive, it is economical to use them in many applications.

Emergency Flasher. The UJT can serve as a triggering circuit for an emergency flasher. See Figure 8-28. UJT Q_1, as the triggering circuit, provides base bias to drive transistors Q_2 and Q_3 through resistors R_2 and R_3. Transistors Q_2 and Q_3 are used to light the incandescent lamp load.

The circuit repetition rate (frequency) is determined by the characteristics of the UJT, the supply voltage, and the emitter *RC* time constant of Q_1. To change the flashing rate, the value of *C* must be changed. As *C* increases in value, the flashing rate decreases. As *C* decreases in value, the flashing increases.

Precision-Temperature Controller. A UJT can serve as a triggering circuit for a precision-temperature controller. See Figure 8-29. Thermistor R_4 is used as a temperature detector for feedback information. UJT Q_1 provides the triggering circuit for the triac. The triac turns the AC power applied to the heater element ON and OFF.

More specifically, full-wave rectified and clamped DC is provided by resistors R_1, R_2, and diodes D_1 through D_5, for driving the UJT. Emitter capacitor *C*

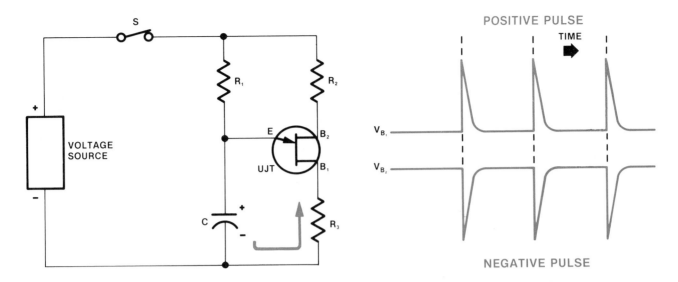

Figure 8-27. With the emitter junction forward biased, the internal resistance of the emitter-B_1 region will drop dramatically and cause capacitor C to discharge its energy through base load resistor R_2. Each time this sequence occurs, a positive pulse appears at B_1 and a negative pulse appears at B_2.

is charged rapidly at the beginning of each half cycle to a *pedestal voltage* (determined by the temperature-sensitive voltage divider R_4 and R_5). After reaching the pedestal voltage, C continues charging slowly. This charging process is shown graphically in Figure 8-29. The curve in the graph is called the ramp. The triac is triggered into conduction by a pulse from UJT Q_1 whenever the UJT emitter voltage on C reaches the UJT triggering level. Raising the pedestal voltage (by cooling the thermistor) causes the ramp to reach triggering level earlier in the cycle, and provides more power to the heater. As more heat goes to the thermistor, the process is reversed, thus providing the regulating function.

Time-Dependent Lamp Dimmer. A lamp dimmer automatically increases or decreases the brightness of a lamp over an adjustable period of time. It can have many functions. See Figure 8-30. Once the OFF function of the switch is activated on the dimmer, the bright lights fade away slowly over a period of 15 to 20 minutes. The delay feature of the dimmer can also eliminate the blinding shock of turning ON a light when a person's eyes are accustomed to darkness.

In the operation of a time-dependent lamp dimmer circuit, the DC voltage for the trigger circuit is derived from zener diodes D_5 and D_6. D_5 and D_6 clamp the pulsating DC voltage (from the full-wave rectifier bridge D_1, D_2, D_3, and D_4) to approximately 15 volts. UJT Q_3 delivers a trigger pulse to the triac. Depending on whether the trigger pulse is delivered late or early

in the cycle, the output to the load (lamp) will vary from completely OFF to completely ON.

The trigger circuit is designed so that a time-dependent output is obtained after initially energizing the circuit. Delay in turning ON or OFF is obtained

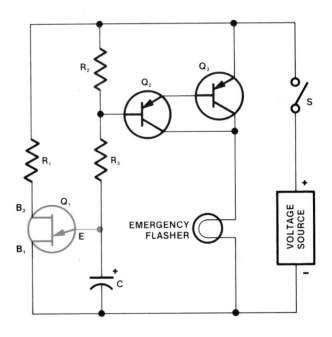

Figure 8-28. UJT Q_1 can serve as a triggering circuit for an emergency flasher.

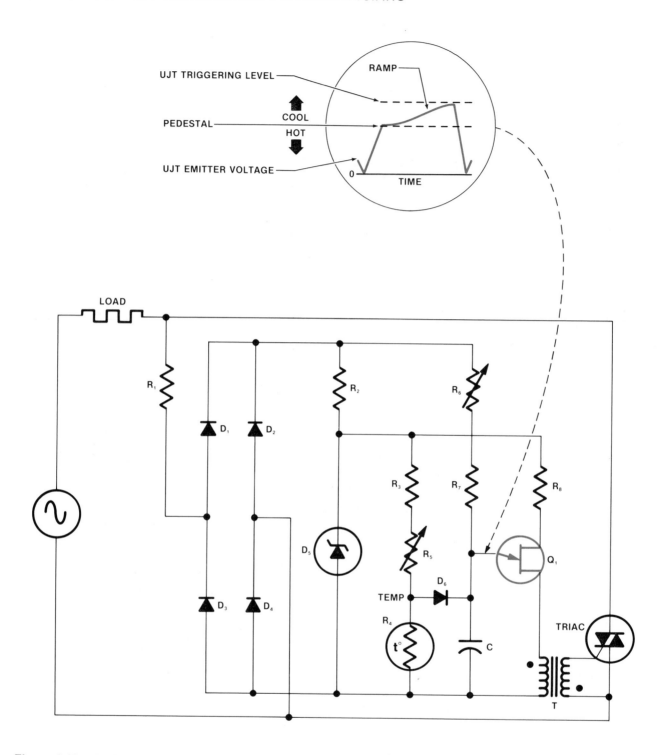

Figure 8-29. UJT Q_1 can serve as a triggering circuit for a precision-temperature controller.

after the position of switch S_1 is changed. When S_1 is placed in the UP position, capacitor C_1 begins to charge through R_2 and R_3. For time periods shortly after switching, C_1 voltage is low. This holds the base current of Q_1 low. Therefore, the emitter of Q_2 is held at a low voltage (below the peak-point voltage on the UJT). Simultaneously, C_2 is charged during each half cycle through R_7. The time constant of R_7-C_2 is long compared to a half cycle of the line voltage. This R_7-C_2 time constant is selected so that the C_2 voltage barely

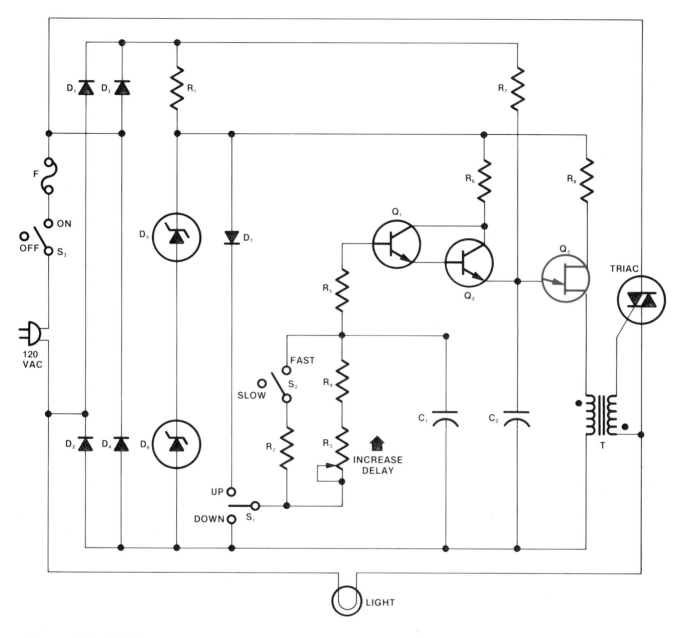

Figure 8-30. UJT Q_3 can serve as a triggering circuit for a time-dependent lamp dimmer.

reaches the peak-point voltage at the end of the half cycle with zero voltage on C_1. As the voltage on C_1 rises, the voltage on C_2 also rises. The R_7-C_2 charging curve then starts from a slightly higher voltage at each cycle. Therefore, the voltage on C_2 reaches the peak-point voltage of the UJT slightly earlier during each cycle, and the output is slowly increased. The double emitter-follower configuration Q_1-Q_2 provides an extremely high impedance so that the charging and discharging currents to C_1 are not shunted away from it.

When switch S_1 is moved to the DOWN position, capacitor C_1 discharges through R_2 and R_3. The operation proceeds as before, but in reverse.

Potentiometer R_3 varies the speed that turns ON or OFF the light. For the circuit of Figure 8-30, the time duration from completely ON to completely OFF, or vice-versa, can be as long as 20 minutes. A bypass switch, S_2, and resistor R_4, provide a method for turning the light ON like a regular switch. S_3 is used to turn ON the device before it can perform the other functions described.

Chapter 8 - Review Questions

1. What is a triac?
2. How are the terminals of a triac designated?
3. Which combined devices most closely resemble the triac?
4. What happens when the voltage on a triac goes too high?
5. What have triacs replaced in control circuits?
6. Why would triacs be useful in explosive environments?
7. What is the procedure for testing a triac?
8. What is the primary use for a diac?
9. Define negative-resistance characteristic.
10. Describe the procedure for testing a diac with an ohmmeter.
11. How is a unijunction transistor (UJT) constructed?
12. How are the leads to a UJT labeled?
13. What is the equivalent circuit of a UJT?
14. What region on a UJT curve does the UJT have its greatest application?
15. What is the primary use for a UJT?

9 TRANSISTOR AS AN AC AMPLIFIER

Transistors can be used as AC amplification devices as well as DC switching devices (discussed in Chapter 6). Amplification is the process of taking a small signal and making it larger. In control systems, transistor AC amplifiers are used to increase small signal currents and voltages so they can do useful work. Amplification is accomplished by using a small input signal to control the energy output from a larger source, such as a power supply.

Key Words

Alpha (α)	Common-emitter amplifier	Linear amplifiers
Bandwidth	Current gain	Narrow bandwidth
Bel	Decibel	Output impedance
Cascaded amplifiers	Distortion	Phase inversion
Class A operation	Emitter-follower circuit	Power gain
Class AB operation	Frequency response curve	Single stage amplifier
Class B operation	Gain	Square wave pulse
Class C operation	Half-power points	Voltage gain
Common-base emitter	Input impedance	Wide bandwidth
Common-collector amplifier	Isolation amplifier	

AMPLIFIER GAIN

The primary objective of the amplifier is to produce *gain*. Gain is a ratio of the amplitude of the output signal to the amplitude of the input signal. In determining gain, the amplifier can be thought of as a "black box." See Figure 9-1. A signal applied to the input of the black box gives the output of the box. Mathematically, gain can be found by dividing output by input:

$$Gain = \frac{Output}{Input}$$

Sometimes, a single amplifier does not provide enough gain to increase the amplitude for the output signal needed. In such a case, two or more amplifiers can be used to obtain the gain required. For example, amplifier A in Figure 9-2 has a gain of 10, and amplifier B has a gain of 10. The total gain of the two amplifiers is 10 × 10, or 100. If the gain of one amplifier were 8 and the other were 9, the total gain would be 72. Amplifiers connected in this manner are called *cascaded amplifiers*. For many amplifiers, gain is in the hundreds, and even thousands.

INPUT   AMPLIFIER   OUTPUT

Figure 9-1. An amplifier can be thought of as a "black box" with an input and an output.

NOTE: Gain is a ratio of output to input and has no unit of measure, such as volts or amps, attached to it. Therefore, the term gain is used to describe current gain, voltage gain, and power gain. In each case, the output is merely being compared to the input.

Computing Current Gain

The *current gain* of an amplifier is the ratio of its output current to its input current. Mathematically, current gain is expressed as:

$$Gain = \frac{I_{OUT}}{I_{IN}}$$

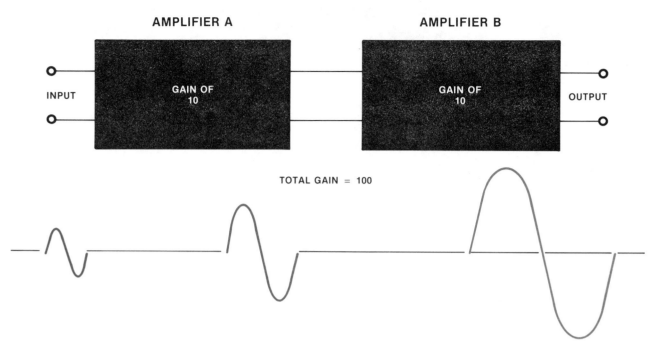

Figure 9-2. Each amplifier comprising the cascaded amplifiers shown has a gain of 10. Combined in this manner, the total gain is 100.

IMPORTANT: Amplifier current gain is not the same as the current gain for an individual transistor, which is shown as beta (β). Beta refers to the transistor as a single device. It does not refer to the entire amplifier circuit containing the transistor.

The transistor current gain (β) is the maximum gain possible for a single-stage (one transistor) amplifier circuit. The actual circuit gain depends on the particular values chosen for load and the type of bias. The amplifier gain for a particular transistor circuit will always be less than the β of the single transistor in the circuit.

Computing Voltage Gain

Voltage gain of an amplifier is the ratio of its output voltage to its input voltage. To calculate voltage gain, measure the RMS output voltage and input voltage of the amplifier, and divide the output by the input. Mathematically, voltage gain is expressed as:

$$Voltage\ gain = \frac{V_{RMS\text{-}OUT}}{V_{RMS\text{-}IN}}$$

Computing Power Gain

Power gain of an amplifier is the ratio of its output power to its input power. Mathematically, power gain is expressed as:

$$Power\ gain = \frac{P_{OUT}}{P_{IN}}$$

BANDWIDTH

The gain of an amplifier is not the same at all frequencies. Amplifiers are designed to operate within a given frequency range. If an amplifier is operated outside of this frequency range, the gain may decrease. The range of frequencies over which the gain of an amplifier is maximum and relatively constant is called its *bandwidth*. The bandwidth for a given amplifier is often shown by a graph called a *frequency-response curve*. Figure 9-3 shows a frequency-response curve for an amplifier that has a working frequency range starting at about 2,000 hertz and stopping at about 20,000 hertz. (This is the range for normal hearing.) The bandwidth for this amplifier is 18,000 hertz, which is the difference between 20,000 hertz and 2,000 hertz. Mathematically, bandwidth is expressed as:

bandwidth = upper frequency − lower frequency

The upper and lower frequency limits are expressed as those points on the frequency-response curve where the gain drops to 0.707 times its maximum gain. These points are called *half-power points* because the power output is one-half the maximum value. Mathematically, this is expressed as:

$$Power\ output = 0.707\ V_{MAX} \times 0.707\ I_{MAX}$$
$$= 0.50\ P_{MAX}$$

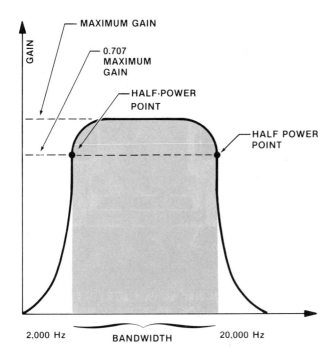

FREQUENCY-RESPONSE CURVE

Figure 9-3. The working frequency of an amplifier can be determined from its frequency response curve. The upper and lower limits are determined by the half-power points. The half-power points exist where the gain drops to 0.707 times the maximum gain.

When the range of frequencies over which an amplifier operates is large, the amplifier has a *wide bandwidth*. When the range is small, the amplifier has a *narrow bandwidth*.

DECIBEL

The human ear responds to the intensity, or loudness, of sound. For example, if an audio amplifier is supplying power to a speaker, the volume of the audio amplifier can be adjusted so that the power delivered to the speaker varies from 1 watt to 100 watts. If a pure sine wave is applied to the input of this audio amplifier and the volume control is adjusted so the power is gradually increased from 1 to 100 watts, it would seem that a gradual increase in volume would be heard. However, this is not the case. If a wattmeter were connected to the load and observed closely, it would show that when the power had increased from 1 watt to approximately 1.26 watts, the ear could detect the increase. As the power is further increased, a change in volume could be detected at approximately 1.58 watts. As the volume is increased again, it could be detected that the power had been increased at

approximately 2 watts, 2.5 watts, 3.17 watts, 3.98 watts, 5.02 watts, and so on, as shown in the following table:

Watts	Watts	Watts
1	5.02	25.1
1.26	6.31	31.7
1.58	7.94	39.8
2	10	50.2
2.5	12.6	63.1
3.17	15.8	79.4
3.98	20	100

This table shows a very important point: The ear is not sensitive to the amount of change, but rather, to the ratio of change that occurs. Notice that as the power changes from 1 watt to 1.26 watts (a .26 watt increase), a change in volume is evident. However, the power has to change from 79.4 watts to 100 watts (an increase of approximately 21 watts) before the change is noticed.

NOTE: The ratio between these two powers is the same:

$$\frac{1.26}{1} = 1.26 \quad \text{and} \quad \frac{100}{79.4} = 1.26$$

Since the human ear responds to the ratio of change rather than to the amount of change, it is convenient to measure the output of audio amplifiers with that in mind. The unit of measure used is the *decibel (dB)*. The decibel is the amount by which a pure sine wave sound must increase before the change in sound level can be distinguished by the average human ear. Each time the power of an amplifier increases, or the loudness of the sound increases, by a ratio resulting in 1.26, it has increased by 1 decibel.

The decibel is a unit of measure used to describe a change between two measurements. In electronics, this change is usually related to voltage or power. The decibel is a subunit of the *bel*. The bel is larger unit than the decibel, and was originally used to show the loss of signal in two miles of telephone wire. Use of the decibel is more common than use of the bel since most changes in small amplifiers are small.

Since decibels denote change, they are often used in describing amplifier gain and line signal losses. For example, an amplifier might have a +30 dB change. The plus sign in front of the number indicates there is an increase between input and output. If a minus sign is in front of the number, as in −10 dB, then there is a reduction in signal between input and output. Figure 9-4 shows decibel gain and loss in relation to a tape-recording system.

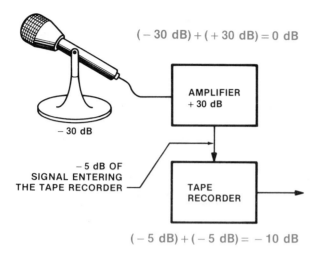

$$(-30\ dB) + (+30\ dB) = 0\ dB$$

AMPLIFIER
+ 30 dB

− 30 dB

− 5 dB OF
SIGNAL ENTERING
THE TAPE RECORDER

TAPE
RECORDER

$$(-5\ dB) + (-5\ dB) = -10\ dB$$

Figure 9-4. A decibel is a unit of measure used to describe a change between two measurements. The positive sign in front of a number indicates that there is an increase between input and output. The negative sign in front of the number indicates a reduction in signal between input and output.

In this system, the microphone and cord provide a loss of − 30 decibels into the amplifier. The amplifier provides a gain of 30 decibels to compensate for the loss. However, the cable and connectors introduce a loss of − 5 decibels prior to going into the tape recorder. If the recorder does not have amplification, it will introduce another loss of − 5 decibels. The net result will be a signal loss of − 10 decibels, which may have to be compensated for by an additional amplifier.

The decibel is widely used in all audio work. Meters are marked in decibels. Amplifiers are rated in decibel gain. Noise levels are measured by the number of decibels they are below the level of the desired sound. Frequency response is graphed in decibels, and microphone outputs are rated in decibels.

TYPES OF TRANSISTOR AMPLIFIERS

The three basic types of transistor amplifiers are the *common-emitter, common-base,* and *common-collector.* See Figure 9-5. The amplifier is named after the transistor connection that is common to both the input and the load. For example, the input of a common-emitter circuit is across the base and emitter, while the load is across the collector and emitter. Thus, the emitter is common to both input and load.

Common-Emitter Amplifier

A transistor amplifies AC through the use of its input circuit and output, or load, circuit. The signal enters the amplifier through the input circuit and exits through the output circuit. See Figure 9-6.

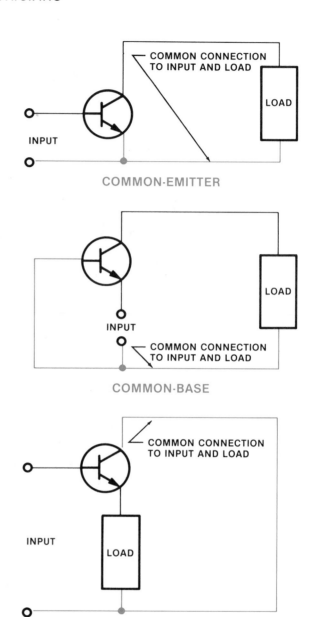

COMMON CONNECTION
TO INPUT AND LOAD

LOAD

INPUT

COMMON-EMITTER

LOAD

INPUT

COMMON CONNECTION
TO INPUT AND LOAD

COMMON-BASE

COMMON CONNECTION
TO INPUT AND LOAD

INPUT

LOAD

COMMON-COLLECTOR

Figure 9-5. The three basic types of transistors and amplifiers are the common-emitter, common-base, and common-collector. The transistor is named after the connection that is common to both the input and the load.

NOTE: The emitter is connected to the ground (reference or common reference point), and both input and output are common to the emitter. Because of this, the common-emitter circuit is sometimes called a grounded emitter circuit. Bias voltage for this circuit is provided by base bias resistor R_B.

Figure 9-7 shows the paths of current in a common-emitter circuit. The value of base resistor R_B and the collector voltage V_{CC} determine the base

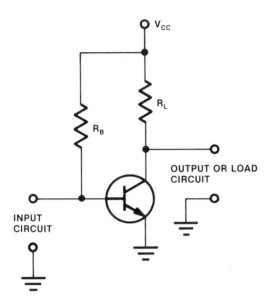

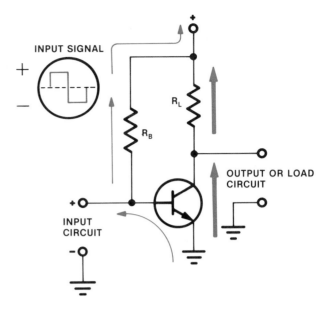

Figure 9-6. The emitter is connected to ground, and the input and output circuits are common to this ground. This circuit is called a common-emitter circuit. It is sometimes also called a grounded emitter circuit because of the common emitter ground.

Figure 9-8. With the square wave pulse applied, the signal rises to a given value and remains at that value for a period of time. It then reverses polarity and rises to a given value until it is time for the cycle to repeat itself. The signal will add to or subtract from the emitter-base bias voltage.

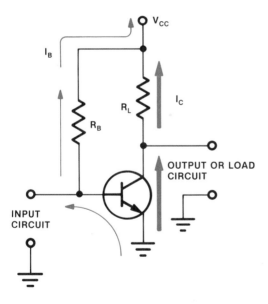

Figure 9-7. The arrows indicate the current paths in a common-emitter circuit. The value of base resistor R_B and the collector voltage V_{CC} determine the base current.

Common-Emitter Amplifier in Dynamic Condition. An input signal must be applied to make an amplifier change from a static condition (at rest) to a dynamic condition (normal operation). See Figure 9-8. In this circuit, a *square wave pulse* is applied to the amplifier. The signal rises to a given value and remains at that value for a period of time. It then reverses polarity and rises to a given value until it is time for the cycle to repeat itself. Because of the alternating nature of the signal, it will add to or subtract from the emitter-base bias voltage.

The input signal increasing in a positive direction adds to the emitter-base forward bias. See Figure 9-9. With an increased forward bias, base current and collector current increase.

The input signal increasing in the negative direction subtracts from, or reduces, the emitter-forward bias. See Figure 9-10. With a decrease in forward bias, base current and collector current decrease.

NOTE: As the input increases, the output signal decreases. As the input decreases, the output signal increases. This is called a phase inversion, *or 180° phase shift.*

Understanding Common-Emitter Phase Inversion. The voltage divider concept is an easy way to explain phase inversion. Consider the output circuit of a transistor as though it were a series voltage-divider network composed of one fixed resistor R_L and one

current I_B. The amount of base current I_B can be found by using the formula:

$$I_B = \frac{V_{CC}}{R_B}$$

variable resistor R_{CE}. It shows the effect of an input signal on the output signal. See Figure 9-11.

In Figure 9-12, top, the total supply voltage V_{CC} equals 10 volts. In this circuit, no signal is applied, and V_{CC} is divided equally across the variable R_{CE} and fixed resistor R_L.

NOTE: These voltages are for illustrating a point and are large in relation to most transistors.

When a negative-going signal is coupled to the base bias, the forward bias is reduced. This causes the collector-emitter section of the transistor to increase in resistance. See Figure 9-12, middle. Since the resistance of R_{CE} has increased in a greater proportion relative to the load R_L, more voltage will be dropped across the output circuit resistance. The output signal V_{CE} will then increase. (In this case, it will increase to 8 volts.)

When the signal is positive, the positive pulse aids the forward bias. This causes the collector-emitter section of the transistor to decrease in resistance. See Figure 9-12, bottom. Since the resistance of R_{CE} now decreases in relation to R_L, less voltage will be dropped across the output circuit resistance. The output signal will then decrease. (In this case, it will decrease to 2 volts.) The output signal of a common-emitter amplifier will be opposite to, or it will be an inversion of, the input signal.

Common-Base Amplifiers

The schematic diagram for the common-base amplifier is shown in Figure 9-13. In this circuit, the input signal is applied to the emitter-base and the output signal is taken from the collector-base. The base is common to both the input and the output. It is the reference point in the circuit and is called ground.

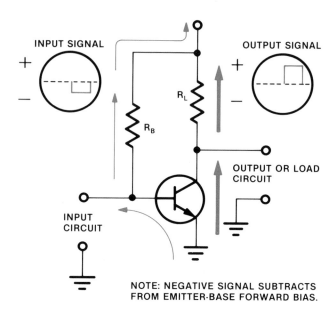

NOTE: NEGATIVE SIGNAL SUBTRACTS FROM EMITTER-BASE FORWARD BIAS.

Figure 9-10. When the input signal increases in the negative direction, it subtracts from the emitter-base forward bias. This causes the base and collector current to decrease.

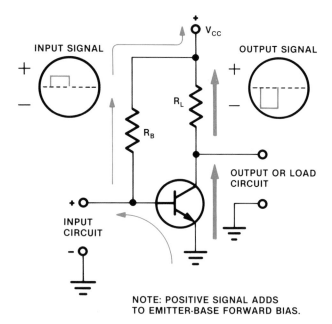

NOTE: POSITIVE SIGNAL ADDS TO EMITTER-BASE FORWARD BIAS.

Figure 9-9. When the input signal increases in a positive direction, it adds to the emitter-base forward bias. This causes the base and collector currents to increase.

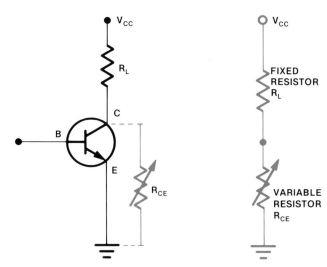

Figure 9-11. As the base current changes, the collector-emitter acts like a variable resistor R_{CE}. When base current increases, resistance decreases. When current decreases, resistance increases.

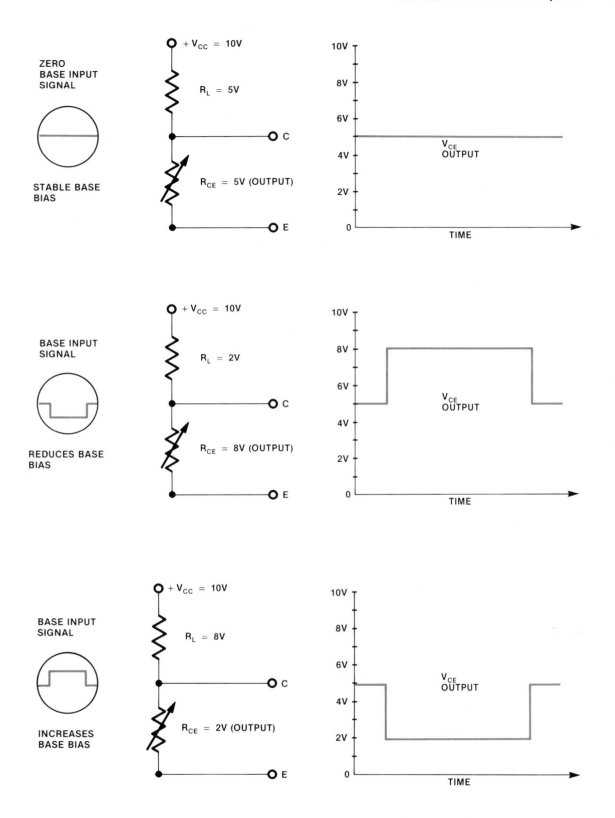

Figure 9-12. TOP: With no signal applied, voltage V_{CC} divides equally across variable resistor R_{CE} and fixed resistor R_L. MIDDLE: With a negative signal, the forward bias is reduced, causing variable resistor R_{CE} to increase. With more resistance, more voltage will be dropped across the output circuit resistance and the output signal will increase. BOTTOM: With a positive signal, forward bias is increased, causing variable resistor R_{CE} to decrease. With less resistance, less voltage will be dropped across the output circuit resistance and the output signal will decrease.

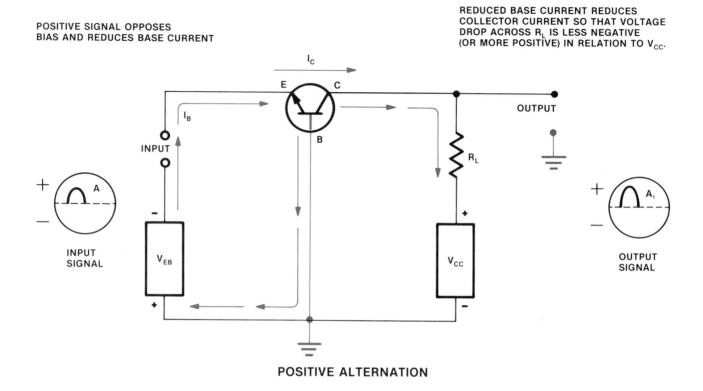

POSITIVE SIGNAL OPPOSES
BIAS AND REDUCES BASE CURRENT

REDUCED BASE CURRENT REDUCES
COLLECTOR CURRENT SO THAT VOLTAGE
DROP ACROSS R_L IS LESS NEGATIVE
(OR MORE POSITIVE) IN RELATION TO V_{CC}.

POSITIVE ALTERNATION

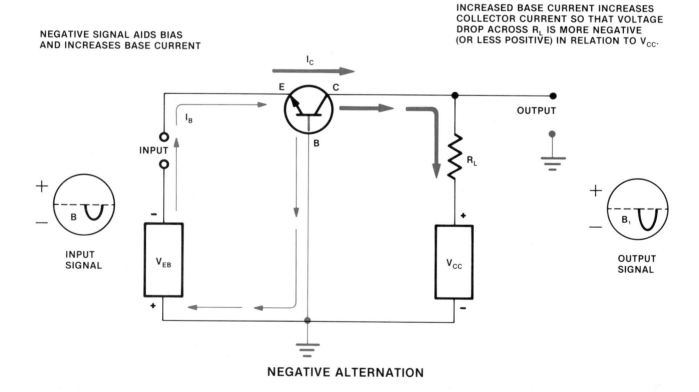

NEGATIVE SIGNAL AIDS BIAS
AND INCREASES BASE CURRENT

INCREASED BASE CURRENT INCREASES
COLLECTOR CURRENT SO THAT VOLTAGE
DROP ACROSS R_L IS MORE NEGATIVE
(OR LESS POSITIVE) IN RELATION TO V_{CC}.

NEGATIVE ALTERNATION

Figure 9-13. In the common-base circuit, the input signal is applied to the emitter-base and the output signal is taken from the collector-base. The base is common to both the input and the output and is called ground.

Common-Base Circuit Operation. On the first alternation of the input signal, when the input signal is positive, it opposes the emitter-base voltage. It also reduces the emitter current accordingly. This reduces the collector current and thus, the voltage drop across output load resistor R_L decreases. The output voltage from collector to ground is equal to the algebraic sum of the voltage drop across R_L and the voltage of the collector source. Therefore, the reduction in voltage across R_L produces an increase in voltage from collector to ground. Thus, the output signal moves in a positive direction as the input signal moves in a positive direction. See Figure 9-13.

Similarly, when the input signal swings in a negative direction, it adds to the emitter-base bias. Adding to the emitter-base bias increases collector current and produces an increased voltage drop across R_L. This increased voltage drop opposes more of the collector source voltage. Therefore, it reduces the output signal from collector to base. As the input signal moves in the negative direction, the output signal moves in the negative direction. See Figure 9-13.

NOTE: In the common-base amplifier, the input signal is applied directly to the emitter. The emitter current controls the collector current.

The amplification factor of all common-base amplifier circuits is designated by *alpha* (*α*), the current gain.

NOTE: This is not the same as beta (β), which is based on a change in base current. Alpha is based on a change in emitter current.

Alpha is the ratio of collector current (I_C) to the emitter current (I_E) with a constant collector voltage. Mathematically, it is expressed as:

$$\alpha = \frac{\Delta I_C}{\Delta I_E}$$

(Alpha is also often called the forward-current transfer ratio.) Since the emitter current is always larger than the collector current in this circuit, the current gain for a common-base circuit is always less than one. Therefore, the common-base circuit is rarely used.

Common-Collector Amplifiers

The schematic diagram for the common-collector amplifier is shown in Figure 9-14. Because the output is taken from the emitter, this circuit is also called an *emitter-follower circuit.* It looks somewhat like the common-emitter circuit. However, the common-collector circuit has the input signal applied between the base and collector, while the output is taken from the collector-emitter.

Biasing circuits for a common-collector amplifier are identical to those for common-emitter circuits.

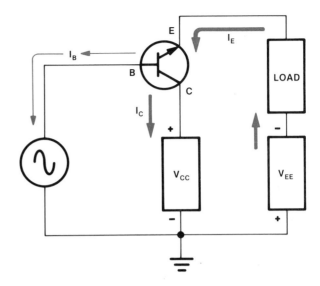

Figure 9-14. In the common-collector circuit, the output is taken from the emitter. The common-collector circuit is also called an emitter-follower circuit.

However, there is one important difference. The difference is that the output voltage for this circuit is equal to the input voltage minus the voltage drop from the base of the transistor to the emitter.

Common-Collector Circuit Operation. In the common-collector circuit, the input signal is applied between the base and collector. The input signal either aids or opposes the forward bias of the transistor. When aiding, base current I_B increases. When opposing, I_B decreases. The change in I_B causes a corresponding change in emitter current I_E and collector current I_C.

The output voltage of the circuit is developed across the load. The load is connected between the emitter and collector regions of the transistor. See Figure 9-14. The emitter current (I_E) flowing through the load is much greater than the base current (I_B). Therefore, the circuit provides an increase in current between the input and output terminals. The voltage developed across the load is always slightly lower than the voltage applied to the circuit. The slightly lower voltage appears at the emitter of the transistor. This is because the transistor tends to maintain a relatively constant voltage drop across its emitter-base junction. The forward voltage drop may be equal to approximately 0.2 volt if the transistor is made of germanium, or 0.6 volt if the transistor is made of silicon. The output voltage appearing at the emitter, therefore, tends to track, or follow, the input voltage applied to the base of the transistor.

The common-collector circuit functions as a current amplifier, but it does not produce an increase in voltage. The increase in output current results in a moderate increase in power. The input resistance of any transistor connected in a common-collector arrangement is extremely high. This is because the input resistance is the resistance that appears across the reverse-biased collector-base junction. The input resistance can be as high as several hundred thousand ohms in a typical low-power transistor. The output resistance appearing between the emitter and collector regions of the transistor will be much lower because of the relatively high emitter current I_E that flows through the output lead. (It is often as low as several hundred ohms.)

The common-collector circuit is used widely in applications where its high input resistance and low output resistance can perform a useful function. The circuit is often used to couple high impedance sources to low impedance loads. Therefore, it can perform the same basic function as an impedance-matching transformer.

NOTE: Since the output signal corresponds to the input signal, there is no phase shift from input to output.

SETTING THE OPERATING POINT ON THE LOAD LINE

A common-emitter amplifier schematic and its corresponding characteristic curves are shown in Figure 9-15. The quiescent point (Q point), or operating point (established on the load line), is where the transistor is biased when no signal is applied to the input.

Operating points should be chosen to accomplish the basic operation of the circuit. The signal swing in Figure 9-16 has been set so that the signal never reaches the actual extremes of saturation or cutoff. Distortion of the signal would result if the signal reaches saturation or cutoff.

Linear Amplification

Figure 9-17 shows an operating point where the input and output signals are linear. *Linear amplifiers* increase the signal and maintain an exact duplicate of the signal being fed in. In other words, the output signal is the same as the input signal, only it is larger. Stereo amplifiers are linear so that the music will be duplicated with the least amount of *distortion*. Distortion is any undesirable change in the original signal.

Signal Change Due to Saturation

Figure 9-18 shows (on a load line) how a signal might be clipped or distorted by an operating point located

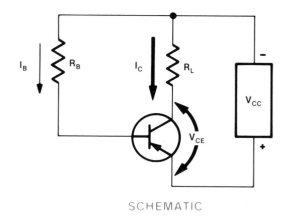

SCHEMATIC

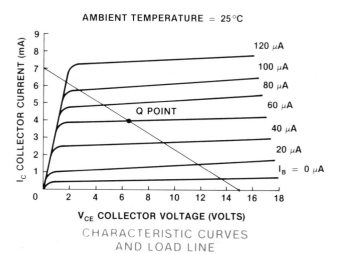

CHARACTERISTIC CURVES
AND LOAD LINE

Figure 9-15. Shown above is a typical common-emitter amplifier schematic, and characteristic curves with Q point already established on the load line.

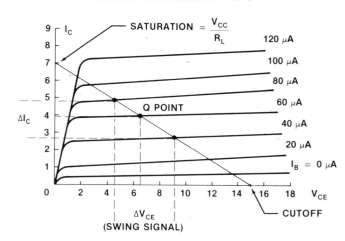

Figure 9-16. When selecting an operating point, the signal swing should never reach the extremes of saturation or cutoff.

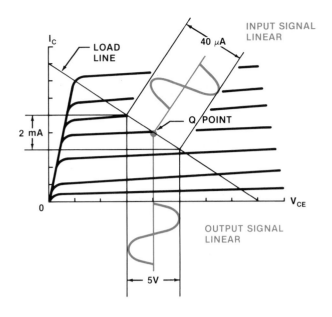

Figure 9-17. Linear amplifiers have an output signal that is the same as the input signal, only it is larger.

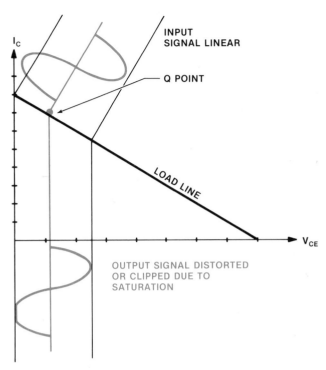

Figure 9-18. A signal may be clipped or distorted by operating at a point too close to saturation.

too close to saturation. When the input signal swings positive, the transistor reaches its maximum output even though the signal continues to increase. There is no more output even with further signal increases. Any increase of signal is lost. In certain cases, clipping may be desirable, but in audio circuits, it is considered a form of distortion.

Signal Change Due to Cutoff

Figure 9-19 shows (on a load line) how a signal may be clipped or distorted by an operating point located too close to cutoff. When the signal swings negative, the signal will continue to be reproduced until the transistor reaches cutoff. Through cutoff, that part of the signal is also lost. Again, clipping of this nature is unacceptable for linear operation. However, it may be used in certain types of circuits, such as a radio transmitter circuit.

Signal Change Due to Saturation and Cutoff

Figure 9-20 shows (on a load line) what happens to a signal on a normal operating amplifier when the input signal is too large. In this case, clipping or distortion occurs on both ends (alternations). A good example of this type of problem is a volume control turned up too high. A positive use of this is to develop switching pulses to a trigger circuit.

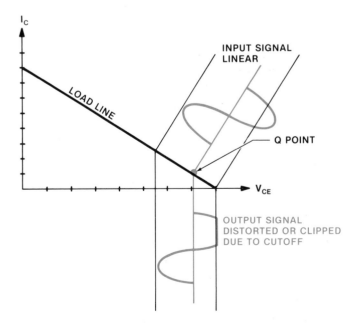

Figure 9-19. A signal may be clipped or distorted by operating at a point too close to cutoff.

CLASSES OF OPERATION

The four main classes of operation for an amplifier are designated by the letters A, B, AB, and C. In each case, the letter is a reference to the level of an amplifier operation in relation to the cutoff condition. The cutoff condition is the point at which all collector current is stopped by the absence of base current.

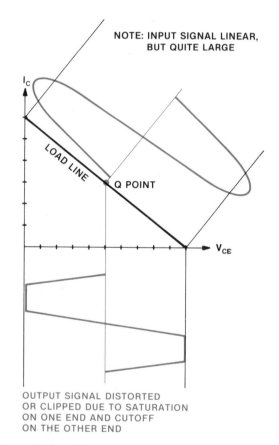

Figure 9-20. When the signal is too large, clipping and distortion may occur at both saturation and cutoff.

Class A Operation

Class A operation is well above cutoff. It is near the center of the straight portion of the characteristic curve on the load line. With class A operation, collector current flows during the entire input signal. The output should duplicate the input and increase it in amplitude. Class A amplifiers are also called linear amplifiers. Figure 9-21 shows the output from a class A amplifier. Class A amplifiers are used as audio amplifiers where true reproduction of the waveform is demanded.

Class B Operation

Class B operation is located near the lower end of the load line closer to cutoff. A class B amplifier reproduces only 180° of the input signal. Collector current flows only during one-half of the input signal. Operation during the negative half cycle produces no base current and no collector current. Figure 9-21 shows the output from a class B amplifier. Class B amplifiers are used mostly with a pair of transistors in the power output of audio amplifiers.

Class AB Operation

Class AB operation is somewhere between class A and class B amplifiers on the load line. With class AB operation, collector current flows for more than half the input signal, but less than the full input signal. Figure 9-21 shows the output from a class AB amplifier.

Class C Operation

Class C operation is located closest to cutoff. With class C operation, collector current flows for less than one-half the input signal. Figure 9-21 shows the output from a class C amplifier. Class C operation is usually found in radio-frequency transmitters.

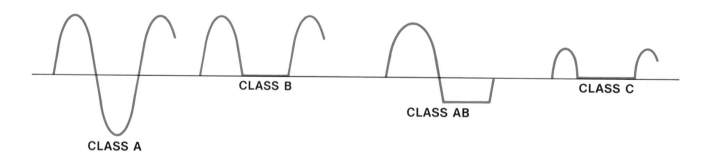

Figure 9-21. With class A operation, collector current flows during the entire input signal. With class B operation, current flows only during one-half of the input signal. With class AB operation, collector current flows for more than half the input signal, but less than the full input signal. With class C operation, collector current flows for less than one-half the input signal.

INPUT AND OUTPUT IMPEDANCES

Figure 9-22 illustrates *input impedance*. Input impedance is the loading effect the amplifier presents to an incoming signal. In other words, the signal sees the amplifier as a load. The amount of load or resistance is the input impedance. Figure 9-23 illustrates *output impedance*. Output impedance is the loading effect the amplifier presents to another device. The device could be another stage of amplification or an output device, such as a speaker. The amount of input impedance and output impedance varies with different types of amplifier configurations. Figure 9-24 shows several common devices which must be impedance-matched to the amplifier.

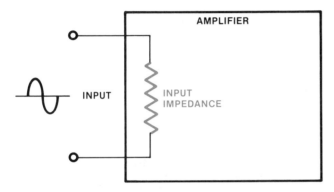

Figure 9-22. Input impedance is the loading effect the amplifier presents to an incoming signal.

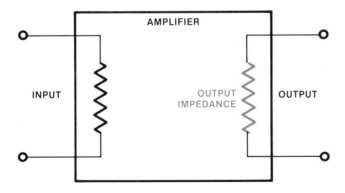

Figure 9-23. Output impedance is the loading effect the amplifier presents to another device.

Input and Output Impedance of a Common-Emitter Amplifier

A typical PN junction has some resistance because of the presence of impurities and minority carriers. This resistance is relatively low (usually a few hundred ohms). It is called the base-emitter impedance, or the input impedance.

NOTE: Base-emitter impedance is a low value and is the input impedance of the common-emitter amplifier.

The reverse-biased collector-base circuit with a relatively large voltage applied to the circuit has very little current flow because of the reverse bias connection. In other words, it appears as a large value of resistance. A typical resistance value for a transistor is 100,000 ohms. The collector-emitter circuit represents the output for a common-emitter amplifier and is a high impedance.

Input and Output Impedance of a Common-Base Amplifier

The amount of change in emitter-base voltage needed to cause an appreciable change in the emitter current is very small. (It is usually 100 ohms or less.) Hence, the input circuit presents a low impedance to the source. However, the output impedance of the transistor is very high (usually exceeding 100,000 ohms). It is high because the collector current is independent of the collector-base voltage. This large ratio of output impedance to input impedance makes it possible for this amplifier to produce a large voltage gain. A large voltage gain is produced despite the absence of current gain. Power gain for this amplifier is moderate.

Input and Output Impedance of a Common-Collector Amplifier

The input impedance of the common-collector circuit is high (approximately 100,000 ohms). The output impedance is low (1,000 ohms or less). The common-collector transistor circuit is used primarily for impedance-matching or for isolation of coupling transistors. The common-collector circuit also has the ability to pass signals in either direction (bilateral operation).

The common-collector circuit is often used as an *isolation amplifier* because of its high input impedance. Its high impedance loads the input signal and allows very little current flow. This in turn causes the signal to be isolated from the other stages. Generally, when a common-collector circuit is used in this way, a common-emitter amplifier follows to provide good gain. See Figure 9-25.

TRANSISTOR SPECIFICATION SHEETS

Semiconductors have their own unique characteristics. The characteristics of these devices are presented on specification sheets. A typical manufacturer's

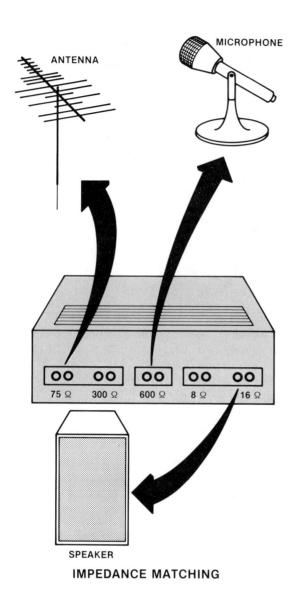

IMPEDANCE MATCHING

Figure 9-24. When connecting microphones, antennas, speakers, tape decks, or other types of devices to an amplifier, each must be impedance-matched for proper operation.

specification sheet for a transistor contains a lead paragraph, maximum ratings, mechanical data, heat sinking information, and characteristic curves.

Lead Paragraph

The lead paragraph of a transistor specification sheet is a general description of the device. The three items the specification sheet usually contains are the kind of transistor, a few major applications, and general features.

Maximum Ratings

The maximum ratings for a transistor must not be exceeded under any operating conditions. Permanent damage to the device would result if the maximum

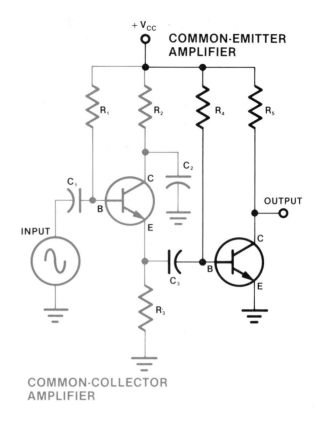

Figure 9-25. The common-collector transistor circuit can be used for impedance-matching or for isolation of coupling stages. When a common collector is used in this manner, a common-emitter amplifier follows to provide good gain.

ratings were exceeded. Typically, transistors are operated well below these ratings to ensure long life and proper performance.

The maximum ratings, as specified in data sheets, have been defined by the Joint Electron Device Engineering Council (JEDEC) and standardized by the Electronics Industries Association (EIA). The maximum ratings include voltage, current, power dissipation, and temperature.

Mechanical Data

Mechanical data usually includes items such as weight, mounting position, dimensions, lead length, and type of case.

Heat Sinking Information

Transistor collector heat dissipation is dependent on ambient temperatures. As the temperature increases, the permissible wattage dissipation decreases. When replacing transistors, the wattage rating at various temperatures is critical and must be taken into consideration for proper heat sinking. Heat sinking information is usually provided in a chart or graph.

Characteristic Curves

Most manufacturers provide a set of characteristic curves on their specification sheets. These characteristic curves can be used in designing and developing new circuits. Generally, the electrician or technician has no need to use these curves.

TRANSISTOR TESTERS

Three types of semiconductor testers are:
1. In-circuit type
2. Field-service type
3. Laboratory-standard type

The in-circuit type of transistor tester is used as a quick method to determine whether or not a transistor is still operating. The primary advantage of this type is that the component does not have to be removed from the circuit. It should be noted that the information gained from this instrument is limited. The same information could be obtained just as readily by using a high impedance voltmeter.

Field-service type of transistor testers give more information about the operating condition of the transistor than the in-circuit type. The field-service type of transistor tester detects current gain, leakage current, and any shorts or opens in a PN junction. Again, it should be noted that with proper voltage measurements and a few simple calculations, all this information is available to the technician.

The laboratory-standard type of tester is found in laboratories and quality assurance stations. Laboratory-standard type of transistor testers give information that voltage checks cannot give. They measure transistor characteristics under actual operating conditions. These types of transistor testers are expensive and are generally not used for service work.

TRANSISTOR SERVICE TIPS

When servicing a transistor circuit, several points should be observed to prevent damage to the transistor:

1. Always be sure that voltage is OFF before removing or installing a transistor. This prevents surge currents from damaging the transistor.

2. Before the voltage is reconnected to a circuit, be sure that the transistor is firmly and correctly inserted into its socket.

3. Because transistors are very sensitive to improper bias voltages, testing them by short circuiting various points to ground should be avoided. A short circuit from base to collector would immediately destroy the transistor.

4. When making voltage measurements, do not allow meter probes to short circuit terminals that are close together.

5. A voltmeter should have a sensitivity rating of 20,000 ohms per volt or more on all ranges when used with transistor circuits. A high-impedance voltmeter, such as an FET meter, is preferred.

6. When making resistance measurements on a transistor, the transistor must be removed from its socket or unsoldered for testing.

7. When bench testing transistor circuits, it is important to use well-regulated power supplies. Semiconductors respond to very minute changes in current. Poorly regulated power supplies may introduce other problems in addition to the one that is being corrected.

Chapter 9 - Review Questions

1. What is amplification?
2. What is gain?
3. Explain the difference between amplifier current gain and beta (β).
4. Define bandwidth.
5. What is a decibel?
6. What are the three basic types of transistor amplifiers?
7. What is phase inversion?
8. What forward voltage drops are usually associated with germanium and silicon?
9. What is another name for a common-emitter circuit?
10. Where are common-collector circuits used?
11. What is a linear amplifier?
12. Define distortion.
13. What causes clipping?
14. What are the four main classes of amplifier operation?
15. What is another name for a class A amplifier?
16. Define input impedance.
17. Define output impedance.

10 FIELD-EFFECT TRANSISTOR AND MULTISTAGE AMPLIFIER

A multistage amplifier is needed to increase a signal if one amplifier does not provide enough gain. In multistage amplifiers, the signal is coupled, or connected, to several stages of amplification. Each stage is a complete amplifier and produces a specific amount of gain. The advantage of this system is the multiplication factor involved. If the first stage has a gain of 40 and the second has a gain of 50, the result is a gain of 40 times 50, or 2000 (not 90).

In addition to gain, a multistage amplifier can provide protection by combining different types of transistors in sequence to obtain certain results. For example, many amplifiers have a very low input impedance. By using a field-effect transistor (FET) on the front end of the multistage amplifier, high impedance can be obtained. The characteristics of FETs make them ideal for preventing excessive loading on the circuit.

Key Words

Amplifier coupling	Enhancement mode	Normally-on MOSFET
Capacitive coupling	Gate	Ohmic region
Common-drain	I_D	P-channel
Common-gate	IGFET	Preamplifier
Common-source	JFET	R_S
Darlington circuit	JK flip-flop	Source
Depletion-enhancement MOSFET	MOSFET	Substrate
Depletion mode	Multistage amplifier	Transformer coupling
Direct coupling	N-channel	V_{DS}
Drain	Normally-off MOSFET	V_{GS}
Dual-gate MOSFET		

FIELD-EFFECT TRANSISTOR (FET)

One type of FET is the junction field-effect transistor (JFET). It is the most common type of FET. The other type of FET is the metal-oxide semiconductor field-effect transistor (MOSFET). The MOSFET is sometimes called an isolated gate field-effect transistor (IGFET) because it has an electrically isolated gate compared to the direct contact of a JFET.

JFET

The two types of JFETs are the *N-channel* and *P-channel*. See Figure 10-1. In both symbols, there are the *gate (G), drain (D),* and *source (S)*. They are found in all FETs. The gate is the control element of a JFET, while the source and drain provide the same function as the emitter and collector on a bipolar transistor. A typical JFET is shown in Figure 10-2.

Uses of JFETs. JFETs may be used as direct current (DC), audio frequency (AF), and radio frequency (RF) amplifiers. They can also be used as switches, gates,

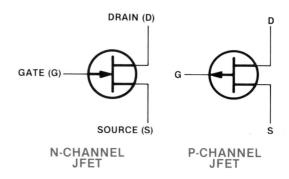

Figure 10-1. The arrow in the N-channel JFET points in on the schematic symbol. On the P-channel JFET, the arrow points outward.

and choppers. JFETs are even found in voltage regulators and current limiters. JFETs are fully compatible with other semiconductor devices, such as standard bipolar transistors, SCRs, triacs, and ICs.

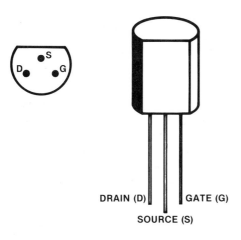

Figure 10-2. A JFET looks much like a bipolar transistor.

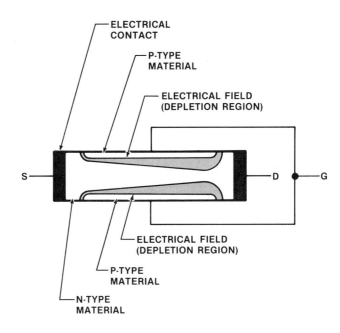

Figure 10-3. The JFET is considered a voltage-driven device. The output current is controlled by an electrical field within it.

JFETs are used in many amplifier configurations because they can amplify a fairly wide range of frequencies, and because they have a high-input impedance. High-input impedance is a definite advantage in the transfer of power. Most bipolar transistors have a low-input impedance and require a relatively powerful input signal to produce a large output at the load. JFETs need very little input power to produce a large output at the load.

Operation of a JFET. A JFET, which is a unipolar device, differs in operation from a bipolar transistor. The JFET is a device in which output current is controlled by the voltage on the input. The input voltage creates an electrical field, or depletion region, within the device. See Figure 10-3. The JFET is considered a voltage-driven device rather than a current-driven device like the bipolar transistor.

The source and drain are connected to a common N-type material. This common material constitutes the channel of the JFET. If a DC potential is connected between the source and the drain, current will flow in the external circuit and through the channel. See Figure 10-4. At zero gate voltage, channel height is maximum and channel resistance is minimum, resulting in maximum current flow. With a slight positive voltage, the channel height opens further, allowing additional current flow.

The two sections of P-type material constitute the gate. Each section has its associated electrical field, or depletion region. If the gate is made negative with respect to the channel, the diodes (formed by the P-type and N-type materials) become reverse biased and the depletion regions increase. At a large enough gate-to-source voltage (V_{GS}), the channel is effectively "pinched off" because the depletion regions touch.

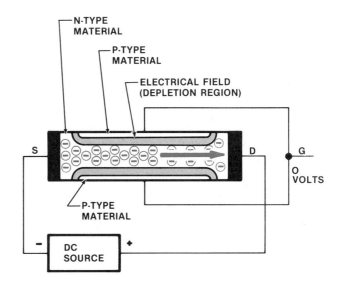

Figure 10-4. With a zero gate voltage, channel height is maximum and channel resistance is minimum, resulting in maximum current flow.

See Figure 10-5. The depletion regions merge at a particular gate-to-source voltage of somewhere between 1 and 8 volts.

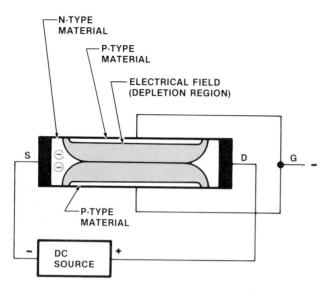

Figure 10-5. At a large enough potential between gate and source (V_{GS}), the channel is effectively pinched off because the electrical fields, or depletion regions, touch each other.

The size of the depletion region is controlled by voltage V_{GS}. When V_{GS} increases, the depletion region increases. When V_{GS} decreases, the depletion region decreases.

V_{GS} controls the *drain current I_D*, and must always provide a reverse bias. This is quite different from a bipolar transistor. In a bipolar transistor, the junction must be forward biased so that the junction impedance is extremely low, and there is current flow in the junction. In a JFET, the junction must be reverse biased so that the junction impedance is high, and there is little current flow in the junction. The major advantage of this is good control using voltage rather than current. The power consumption of the JFET (in standby operation) is thousands of times less than that of a bipolar transistor controlling the same function.

JFET Output Characteristic Curves. The *drain source voltage V_{DS}* is plotted along the horizontal axis, and the drain current I_D is plotted along the vertical axis on the characteristic curve for a typical N-channel JFET. See Figure 10-6. By using one curve, the detailed information produced may be more readily obtained.

The first portion of the curve where the drain current rises rapidly is called the *ohmic region* of the JFET. In this region, the drain current is controlled by the drain-source voltage and the resistance of the channel. On the flattened portion of the graph, the JFET is at the saturation region. In this region, the drain current is controlled primarily by the width of the channel. When the current I_D begins another

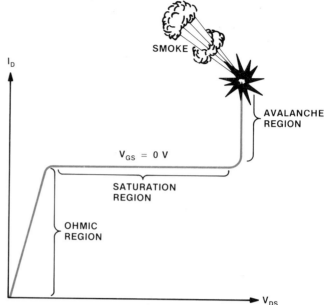

Figure 10-6. The first portion of the JFET characteristic curve is the ohmic region. It is controlled by the drain source voltage V_{DS} and the resistance of the channel. The flattened portion of the graph is the saturation region, and is controlled by the width of the channel. The remaining sharp increase is the avalanche region. It is controlled only by the external resistance of the load.

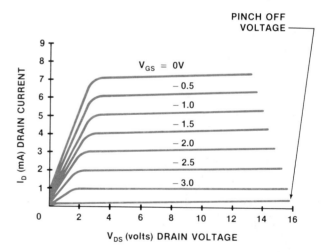

Figure 10-7. A set of curves shows the overall operation of a JFET.

sharp increase, avalanche, or breakdown, begins. At this point, a large amount of current begins to pass through the channel. If this current is not limited by an external resistance or load, the JFET will be damaged. A set of curves shows the overall operation of the JFET. See Figure 10-7.

NOTE: A JFET must never be forward biased. If the voltage applied to the gate source were forward biased, the drain current would increase to very high levels. Also, forward biasing the gate source causes the input impedance to drop to a low value.

Developing the Load Line. A load line drawn on the JFET characteristic curves shows the dynamic operating characteristics. See Figure 10-8. The load line is shown extending from 20 volts (V_{DS}) on the horizontal axis to 20 milliamps (I_D) on the vertical axis.

V_{DS} is determined by using the value for drain voltage at zero drain current I_D or pinch off. The value of I_D is the current value present if the JFET is saturated.

With a bias voltage V_{GS} of -1.5 volts applied to the gate, the voltage is between the JFET and the drain load resistor R_L. One-half (10 volts) of the source voltage is present across the load resistor, and one-half (10 volts) of the voltage is across the JFET. These values of voltage and current exist only for a V_{GS} of -1.5 volts. For other values of gate bias, the drain-to-source voltage and drain current will vary.

The load line is used to show what happens to the output voltage and current when the bias voltage V_{GS} is changed. A change in the gate-to-channel bias voltage can be produced by changing the bias supply voltage. The change can also be produced by applying a signal to the gate and source terminals. Figure 10-9 shows a circuit in which the voltage V_{DD} distributes according to the load resistor R_L and the drain source.

JFET Power Dissipation. The JFET, like most semiconductor devices, can dissipate only a certain amount of power. The safe dissipation limits for a JFET are listed on the manufacturer's specification sheet, which usually has a power dissipation curve.

Normal dissipation rating applies to ambient temperatures of 25 °C or less. If the JFET must operate at a high temperature, it must be derated according to the manufacturer's specifications.

JFET Circuit Configurations

The JFET can be connected into three basic configurations: *common-source, common-gate,* and *common-drain.* The common-source connections for the N-channel and P-channel JFETs are shown in Figure 10-10. The only difference between an N-channel and P-channel configuration is the direction of electron flow in the external circuit, and the polarity of the bias voltages. Common-gate and common-drain are constructed in a similar common element arrangement.

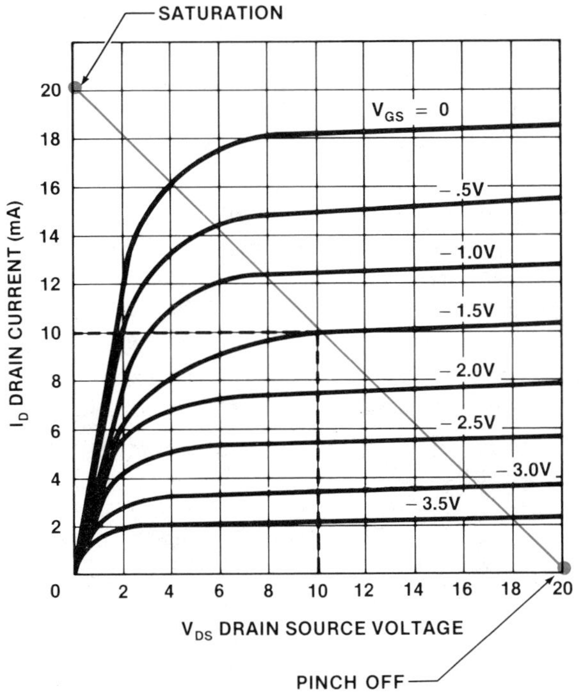

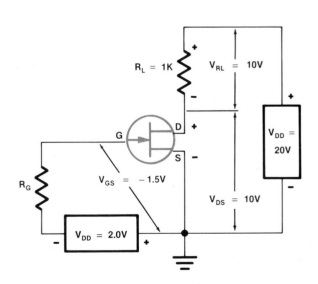

Figure 10-8. A load line drawn on a characteristic curve shows the dynamic operating characteristic of the JFET.

Figure 10-9. Depending upon the value of voltage V_{GS}, voltage V_{DD} distributes accordingly across the load resistor R_L and the drain-source of the JFET. In this case, a V_{GS} of -1.5 volts results in an even distribution of 5 volts across each device. For other values of gate bias, the drain-to-source voltage and drain current will vary.

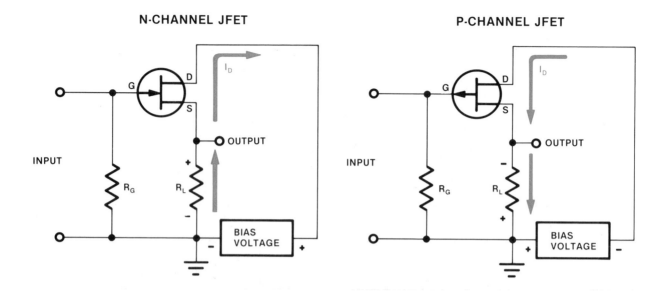

Figure 10-10. In the common-source circuit, the input is across the gate and source terminals and the output is across the drain and source terminals. The source is the common element. The only difference between an N-channel and a P-channel JFET is the direction of electron flow in the external circuit and the polarity of the bias voltage.

Common-Source Amplifier. The common-source JFET can be biased from the voltage drop across a resistor R_S, in series with the source terminal. See Figure 10-11. In each circuit configuration (N-channel and P-channel JFET), the gate bias voltages V_{GS} have opposite polarities and opposite directions of current flow. For the N-channel JFET, the current flow is from ground through *source resistor R_S*. This makes the source side of R_S positive with respect to ground. Since gate resistor R_G is also returned to ground, the voltage across R_G has a polarity that correctly biases the gate negative with respect to the source. The opposite relationship exists in the P-channel JFET. In this circuit, the current direction is such that the source side of resistor R_S is negative with respect to ground. Hence, the gate is made positive with respect to the source. This is the correct polarity for reverse biasing the junction when the channel is made of P-type semiconductor material.

Operation of Common-Source Amplifier. Figure 10-12 illustrates the operation of a common-source amplifier. The AC signal is applied across gate resistor R_1. The DC bias is set by the DC voltage drop across source resistor R_2. To reduce the effect of AC on the DC bias, capacitor C is used as a bypass to shunt any AC signals around R_2. Thus, a constant DC gate-to-source voltage level is maintained. With the input

signal applied across R_1, the signal varies the gate voltage around the DC operating bias, causing a variation in the drain current I_D.

In a P-channel JFET circuit, the positive swing of the input signal increases the gate bias. I_D decreases and the drain voltage becomes more negative, producing a negative swing in the output signal. A negative swing of the input signal decreases the gate bias. The drain current decreases and the drain voltage becomes more positive, producing a positive swing in the output signal.

In the configuration of an N-channel amplifier, a positive swing of the input signal decreases the gate bias. This causes an increase in drain current I_D and a less positive (more negative) swing of drain voltage and output signal. See Figure 10-13. The result is a rise in the positive drain voltage, and a positive swing of the output signal. The output signal at the drain then is 180° out of phase with the input signal. The common-source circuit is the only JFET configuration that produces an inverted output signal.

The input resistance of a JFET can be greater than 100 megaohms. The gate resistor can greatly influence the input resistance. In most cases, the value of the gate resistor determines the input impedance of the stage. Therefore, the gate resistor usually has a very high resistance (1 megaohm or greater).

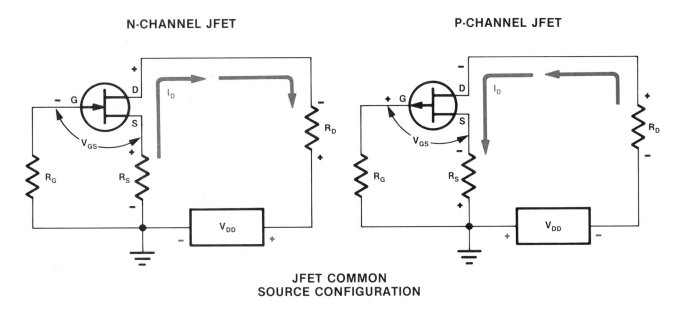

**JFET COMMON
SOURCE CONFIGURATION**

Figure 10-11. The common-source JFET can be biased from the voltage drop across resistor R_S, in series with the source terminal. For the N-channel JFET, the current flow is from ground through source resistor R_S. This makes the source side of R_S positive relative to ground. Since the gate resistor R_G is also returned to ground, the voltage across R_G has a polarity that correctly biases the gate negative with respect to the source. The opposite relationship exists for the P-channel JFET.

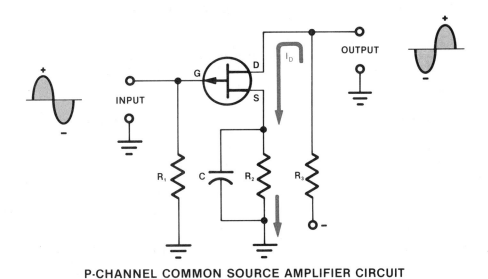

P-CHANNEL COMMON SOURCE AMPLIFIER CIRCUIT

Figure 10-12. In a P-channel JFET circuit, the positive swing of the input signal increases the gate bias. The drain current decreases and the drain voltage becomes more negative, producing a negative swing in the output signal. A negative swing of the input signal decreases the gate bias. The drain current decreases and the drain voltage becomes more positive, producing a positive swing in the output signal.

Common-Gate Amplifier. Figure 10-14 shows two common-gate amplifier circuits using N-channel and P-channel JFETs. In these circuits, the AC input signal is applied to the source and the gate, with the gate at ground potential. The input impedance of the common-gate configuration is the lowest of the three basic configurations. The output signal is present between the drain and the gate, and is a high output impedance.

The common-gate circuit is used in applications where the source of signal requires a low-to-high impedance match. The common-gate is often used in communication circuits. This is because it is capable of good voltage gain and reasonable power gain over a wide range of frequencies (particularly those in the higher frequency ranges). Thus, the common-gate is often found in UHF and VHF amplifiers.

Operation of the Common-Gate Amplifier. The input signal and output signal are in phase for the common-gate amplifier. For the N-channel, a negative swing of the input signal decreases the gate source bias, increases the drain current, and decreases the drain voltage. See Figure 10-14, left. Conversely, a rise in the input signal voltage increases the gate bias, decreases the drain current, and increases the drain and

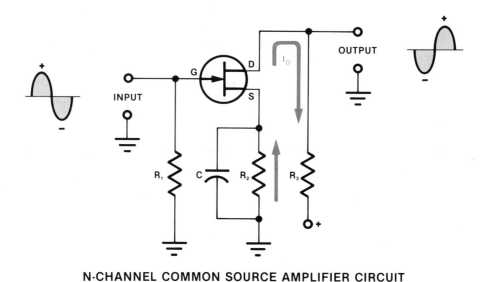

N-CHANNEL COMMON SOURCE AMPLIFIER CIRCUIT

Figure 10-13. In the N-channel amplifier, a positive swing of the input signal decreases the gate bias, causing an increase in the drain current. It also causes a less positive (more negative) swing of drain voltage and output signal. This results in a rise of the positive drain voltage and a positive swing of the output signal.

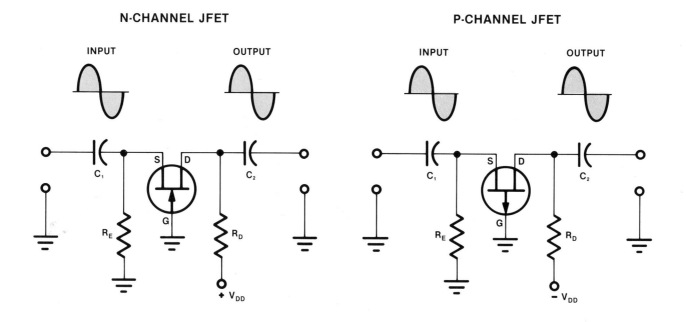

JFET COMMON GATE CONFIGURATIONS

Figure 10-14. The input signal and the output signal are in phase for the common-gate amplifier for both the N-channel JFET and the P-channel JFET.

output voltages. The output voltage is amplified, and is in phase with the input voltage.

In the P-channel, the positive swing of the input signal decreases the gate bias. See Figure 10-14, right. Therefore, the drain current increases and the drain voltage becomes less negative (positive swing of the output signal). Conversely, with the negative swing of the input signal, there is a higher gate-to-source bias. Also, the drain current drops, and the drain voltage becomes more negative.

Common-Drain Amplifier. The basic common-drain circuits for the N-channel and P-channel FETs are shown in Figure 10-15. The input resistance of a common-drain circuit is high. It is high because of the degenerative effect of the AC output voltage developed across source resistor R_S. Like the emitter-follower stage, the output voltage from the common-drain is always less than the input voltage. Also, the circuit is not capable of voltage gain. However, the common-drain does produce good current gain and power gain.

The output impedance is low for the common-drain circuit because the signal is taken from the source terminal. Since the output impedance is low, the common-drain stage is often used when a high-impedance signal source must be matched to a low-impedance load.

Operation of a Common-Drain Amplifier. In the N-channel circuit, a positive swing of the input signal

decreases the gate bias. See Figure 10-15, left. As a result, the drain-to-source current rises, and the source and output voltages swing more positive. Conversely, a negative swing of the input signal increases the gate bias and decreases the drain-to-source current. The source voltage becomes less positive because the current decreases in the source load resistance, and the output voltage swings negative.

For the P-channel circuit, the positive swing of the input signal increases the gate bias. See Figure 10-15, right. Therefore, the drain-to-source current decreases. Also, the voltage drop across the source resistor is less negative (positive swing of the output signal). During the negative swing of the input signal, the gate bias decreases. Therefore, the drain-to-source current rises in the source resistor. Because of the direction of the current, this rise makes the source more negative, and there is a negative swing of the output signal voltage.

The voltage change at the output has the same direction as the input change, so the signals are in phase. Both common-gate and common-drain circuits have in-phase signal outputs. The common-source amplifier is the only JFET configuration that inverts the phase of the signal.

JFET Application Circuits

The JFET is used extensively in circuits where low power and high impedance is a factor. JFETs need very little power to produce a large output at the load.

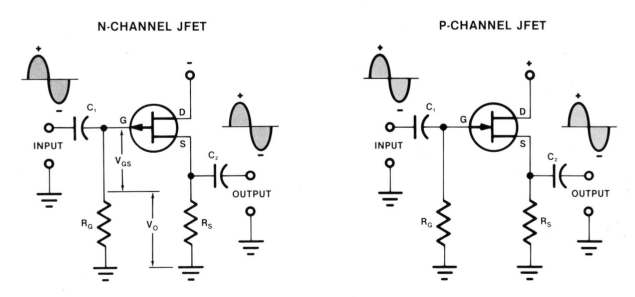

N-CHANNEL JFET   **P-CHANNEL JFET**

JFET COMMON DRAIN CONFIGURATIONS

Figure 10-15. In the N-channel circuit, a positive swing of the input signal decreases the gate bias. Conversely, the negative swing of the input signal increases the gate bias and decreases the drain-to-source current. For the P-channel circuit, the positive swing of the input signal increases the gate bias. During the negative swing of the input signal, the gate bias decreases. The output of both amplifier circuits are in phase with the input.

JFETs are also easily matched to other semiconductor devices, such as SCRs, triacs, and ICs.

Preamplifier (Common-Source). The JFET common-source amplifier can be used as a low-level *preamplifier*. (A preamplifier is a circuit that provides gain for a very weak signal before it goes through the normal stages of amplification.) See Figure 10-16. This circuit permits direct input from a high-impedance, small-signal device, such as a crystal microphone. It then matches the microphone to a low-impedance power amplifier. Source resistor R_S provides gate bias. To obtain a higher AC gain from this circuit, a bypass capacitor C_1 connected across R_S maintains the bias on the gate at the desired operating point.

Impedance-Matching (Common-Drain). Figure 10-17 shows a circuit stepping-down impedance levels while preserving the bandwidth and linearity of the signals being amplified. In this case, a high-impedance microphone is coupled to a low-impedance *coaxial cable* to avoid losing frequency response. The common-drain JFET circuit (Q_1) can also step-up the input impedance of the bipolar transistor Q_2. See Figure 10-18. In effect, the common-drain serves as an impedance transformer with power gain.

JFETs in Multimeters. *FET multimeters* use JFETs to provide high input impedances with good sensitivity. (The J is usually dropped when referring to an FET multimeter.) The JFET is capable of producing an

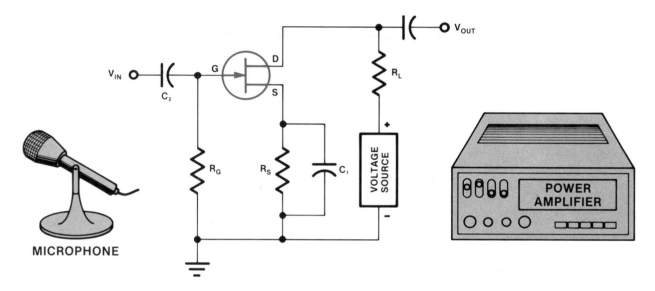

Figure 10-16.   When the JFET is used as a common-source preamplifier, the circuit permits direct input from a high-impedance, small-signal device, such as a microphone, to a low-impedance power amplifier.

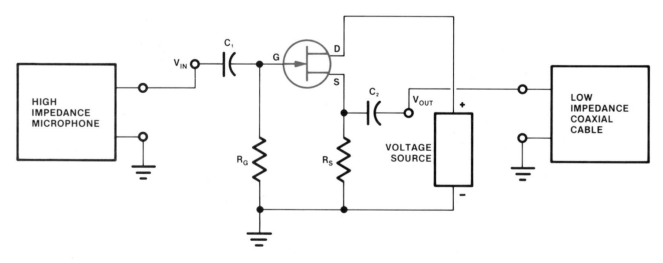

Figure 10-17.   The common-drain JFET circuit couples the microphone to the coaxial cable to preserve bandwidth and linearity.

input of 11 megaohms or more with sensitivities exceeding those of a vacuum tube voltmeter (VTVM).

Figure 10-19 shows a simple FET multimeter circuit. In this circuit, the JFET is one leg in the DC bridge. Source resistor R_S provides negative feedback for high linearity in the response. The circuit requires good regulation, which is provided by the zener diode.

Voice-Operated Relay. In the circuit of Figure 10-20, the JFET provides a high-impedance match of the microphone to the low-impedance of the bipolar transistor Q_2. When the microphone receives a signal, it is amplified through the JFET and the bipolar transistor. It is then applied to the *JK flip-flop*. The JK flip-flop has the unique feature of changing its state when a voltage, or high state, is applied to its J and K inputs. Thus, as each amplified pulse hits the JK flip-flop, the JK flip-flop will either be high or low. If the JK flip-flop is high, the relay will close because transistor Q_3 is turned ON. If the JK flip-flop is low, the relay will drop out because Q_3 is turned OFF.

MOSFET

MOSFETs (metal-oxide semiconductor field-effect transistors) are either three-terminal devices or four-terminal devices. They have the same terminal designations as the JFET (gate, drain, and source). In addition, the four-terminal MOSFET has a designation for the substrate.

The gate voltage of the MOSFET controls the drain current just as it does in the JFET. The impedance of the MOSFET is high. The main operating difference between the JFET and MOSFET is that the MOSFET can have a positive voltage applied to the gate and still have zero gate current. A MOSFET is classified by its mode of operation. It operates either in the *depletion mode* or *enhancement mode*. (See Operation of Enhancement MOSFET and Operation of Depletion-Enhancement MOSFET for explanations.)

MOSFET Schematic Symbols and Lead Identification. MOSFETs are available as N-channel and P-channel devices. See Figure 10-21. On an N-channel

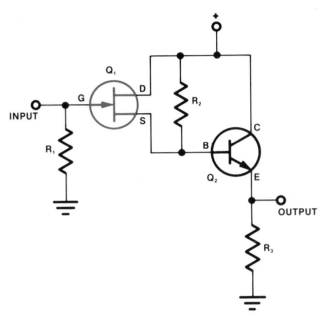

Figure 10-18. The common-drain JFET circuit can step-up the input impedance of a bipolar transistor.

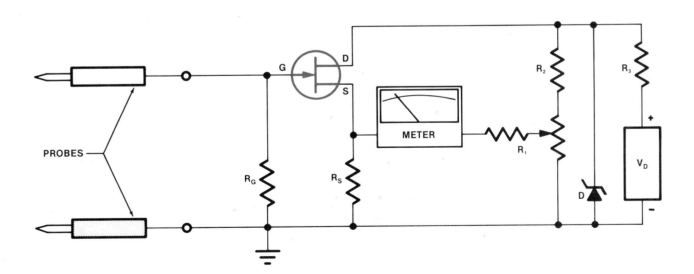

Figure 10-19. A multimeter can provide high-input impedance with good sensitivity by using a JFET in its front-end bridge circuit. In most cases, a JFET multimeter can produce an input of 11 megaohms of more.

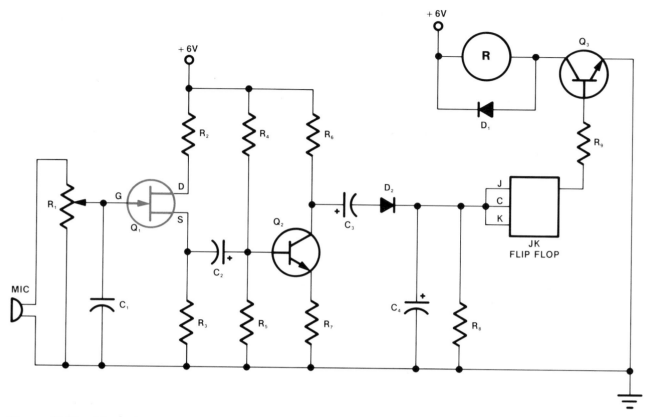

Figure 10-20. The voice-operated relay uses the JFET as a high-impedance match between the microphone and the bipolar transistor. The JK flip-flop in this circuit allows the transistor to be turned ON and OFF, based on inputs from the microphone. This in turn, causes the relay to pull in and drop out. Almost any type of load may be attached to the relay circuit for control.

MOSFET, the arrow on the substrate points toward the channel. With a P-channel MOSFET, the arrow on the substrate points away from the channel.

NOTE: On schematic symbols for four-terminal MOSFETS, the substrate is not connected to the source.

Enhancement MOSFET Construction. The main body of the enhancement MOSFET is composed of a highly resistive P-type *substrate* (material). See Figure 10-22. Two low-resistance N-type regions are diffused in the P-type substrate, forming the source and drain. See Figure 10-23. When the source and drain are completely diffused, the surface of the MOSFET is covered with a layer of insulating material, such as silicon dioxide. See Figure 10-24. Holes are cut into the insulating material, allowing contact with the N-type regions, thereby connecting the source and drain leads. See Figure 10-25. The MOSFET construction is complete when a metal contact (gate) is placed over the insulating material in a position to cover the channel from source to drain. See Figure 10-26.

NOTE: The metal contact is placed on top of the insulating material. There is no physical contact between the gate and the P-type substrate. A MOSFET is sometimes called an insulated (isolated) gate FET, or IGFET, because of this insulation.

Operation of Enhancement MOSFET. A review of capacitor operation helps explain enhancement operation. The metal contact of the gate, the insulating material, and the P-type substrate are essentially a capacitor. The metal gate and the P-type substrate can be considered the plates of a capacitor. The oxide insulator is its dielectric. See Figure 10-27.

A voltage applied to the plates of a capacitor distorts the electrons that are in orbit in the dielectric. See Figure 10-28. In this case, a positive voltage is applied to the upper plate, and the electrons move in the direction of the positive plate.

If this principle is applied to the enhancement MOSFET, the gate can be used to produce a conductive channel from source to drain. The positive charge of a positive voltage placed on the gate induces a negative charge on the P-type substrate. With

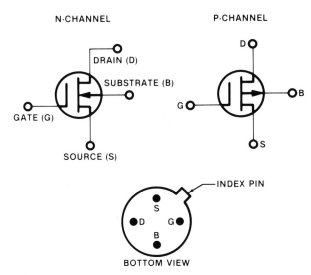

N-CHANNEL   P-CHANNEL

DRAIN (D)

SUBSTRATE (B)

GATE (G)

SOURCE (S)

INDEX PIN

BOTTOM VIEW

FOUR-TERMINAL MOSFET

THREE-TERMINAL MOSFET

Figure 10-21. MOSFETs are available in N-channel and P-channel constructions. The substrate is not connected to the source on a four-terminal MOSFET.

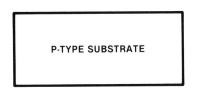

P-TYPE SUBSTRATE

Figure 10-22. The main body of the enhancement MOSFET is composed of a highly resistive P-type substrate.

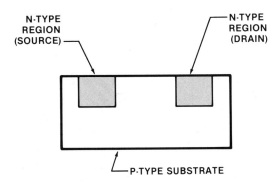

N-TYPE REGION (SOURCE)

N-TYPE REGION (DRAIN)

P-TYPE SUBSTRATE

Figure 10-23. Into the P-type substrate, two low-resistance N-type regions are diffused, forming the source and drain.

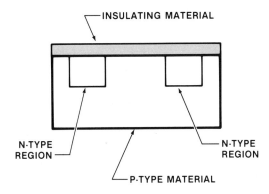

INSULATING MATERIAL

N-TYPE REGION

N-TYPE REGION

P-TYPE MATERIAL

Figure 10-24. When the source and drain are completely diffused, the surface of the MOSFET is covered with a layer of insulating material, such as silicon dioxide.

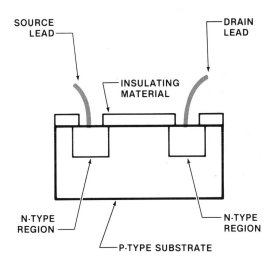

SOURCE LEAD

DRAIN LEAD

INSULATING MATERIAL

N-TYPE REGION

N-TYPE REGION

P-TYPE SUBSTRATE

Figure 10-25. Holes are cut into the insulating material, allowing contact with the N-type regions, thereby connecting the source and drain leads.

increasing positive voltage, the holes in the P-type substrate are repelled until the region between the source and drain becomes an N-channel. See Figure 10-29. Once the N-channel is forward biased between source and gate, current begins to flow.

Since electrons have been added to form the N-channel, this MOSFET has enhanced current flow resulting from the application of a positive gate voltage. With a more positive gate voltage, the channel becomes wider. Therefore, current flows from source to drain due to the decreased channel resistance.

When the gate has a zero voltage or a negative voltage, no enhancement effect is possible and the

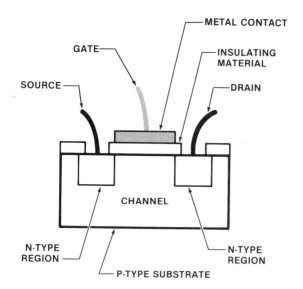

Figure 10-26. The MOSFET construction is complete when the metal contact of the gate is placed over the insulating material to cover the channel from source to drain.

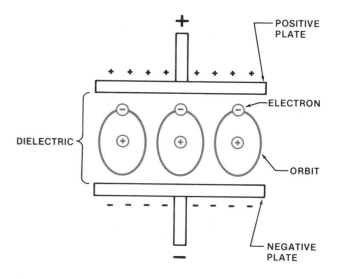

Figure 10-28. A positive voltage applied to the upper plate of a capacitor causes the dielectric to become distorted, with the electrons moving in the direction of the positive plate.

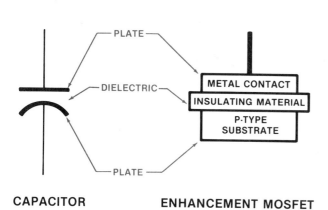

CAPACITOR   **ENHANCEMENT MOSFET**

Figure 10-27. The metal contact of the gate, the insulating material, and the substrate are essentially a capacitor.

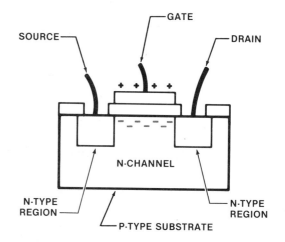

Figure 10-29. The capacitance principle applied to the enhancement MOSFET allows the gate to be used for producing an N-channel from source to drain.

MOSFET will not conduct. Since the enhancement MOSFET does not conduct at zero gate voltage, it is often called a normally-off MOSFET.

NOTE: A P-channel enhancement MOSFET is constructed like an N-channel device, only all P and N regions are reversed. For the P-channel device, a negative gate voltage induces a P-channel, and enhances current flow through the use of holes in the channel.

In the schematic symbol, a broken line between terminals indicates the channel. The broken line signifies that the MOSFET is normally-off. See Figure 10-30.

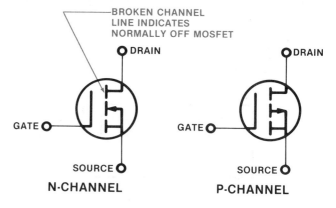

N-CHANNEL   **P-CHANNEL**

Figure 10-30. The broken line indicates that there is no channel established between source and drain when voltage V_{GS} is zero.

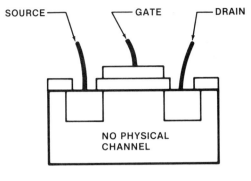

ENHANCEMENT MOSFET

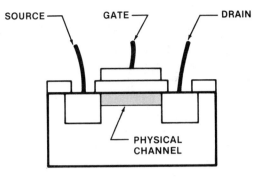

DEPLETION-ENHANCEMENT MOSFET

Figure 10-31. The enhancement MOSFET and the depletion-enhancement MOSFET are very similar. The major difference is the addition of a physical conducting channel between the source and the drain in the depletion-enhancement MOSFET.

Depletion-Enhancement MOSFET. The depletion-enhancement MOSFET is constructed much like the enhancement MOSFET. The major difference is the addition of a physical conducting channel between the source and drain. See Figure 10-31. The presence of the channel allows current to flow from source to drain even without a gate voltage.

Operation of Depletion-Enhancement MOSFET. The depletion-enhancement MOSFET has the same capacitive effect that the enhancement MOSFET has. However, when a negative voltage is applied to the gate, holes from the P-type substrate are attracted into the N-channel. The holes neutralize the free electrons. See Figure 10-32. The result is that the N-channel is depleted, or reduced, in the number of carriers. The depletion of carriers increases channel resistance and reduces current. The operation of a MOSFET with a negative gate voltage is called the *depletion mode*.

With a positive gate voltage, the N-channel MOSFET can operate in the *enhancement mode*. See Figure 10-33. The positive gate voltage widens the N-channel, causing an increase in channel current. Since the depletion-enhancement MOSFET will

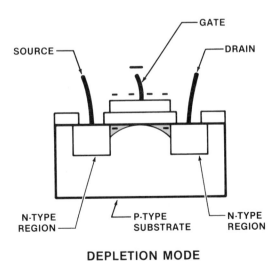

DEPLETION MODE

ENHANCEMENT MODE

Figure 10-32. The depletion-enhancement MOSFET also operates on the capacitive effect. When a negative voltage is applied, the holes from the P-type substrate are attracted into the N-channel, depleting the N-channel of electrons. This depletion increases channel resistance and reduces current.

Figure 10-33. When a positive voltage is applied to the gate, the N-channel widens, causing an increase in channel current. Since the depletion-enhancement MOSFET conducts a significant current even when V_{GS} is zero, it is often called a normally-on MOSFET.

conduct a significant current even when V_{GS} is zero, it is often called a normally-on MOSFET.

Characteristic Curves for Depletion-Enhancement MOSFET. A typical set of characteristic curves for a MOSFET shows the operating ranges for the enhancement and depletion modes. The starting point is at $V_{GS} = 0$ V. See Figure 10-34. When the voltage falls below zero and into the negative voltage region, the MOSFET operates in the depletion mode. When the voltage increases above zero and into the positive voltage region, the MOSFET operates in the enhancement mode. Because the MOSFET can operate in either the depletion mode or the enhancement mode, it is called a *depletion-enhancement MOSFET*.

Dual-Gate MOSFET. MOSFETs are also constructed with two gates. Figure 10-35 shows a *dual-gate MOSFET* arrangement. Current through the MOSFET can be cut off by either gate. It also operates on the capacitive effect.

The dual-gate arrangement allows this MOSFET to be used in a variety of circuits. For example, in a gain control circuit, the audio signal is applied to Gate 1, and the gain control voltage is applied to Gate 2. The gain control voltage can then be used to control the output from Gate 2. The schematic symbol for the dual-gate MOSFET is shown in Figure 10-36.

Installation and Removal of MOSFETS. Care must be exercised when handling a MOSFET since the gate insulation is very thin. Any static charge introduced at the gate can perforate the insulation and destroy the device. Manufacturers of some MOSFETs wrap them in metal foil or short their leads with a metal eyelet or spring for protection. The shorting eyelet should be removed only after the device is installed in its circuit. If a MOSFET is removed from a circuit, it should be wrapped in foil or the leads should be shorted for protection.

MULTISTAGE AMPLIFIERS

A single stage of amplification is often not sufficient to drive a load to the required amount from any given input source. Therefore, various stages of amplification must be connected, or coupled together, to build up the output to the required level. Coupling transfers the signal from the output of one stage to the input of the next without distortion or loss.

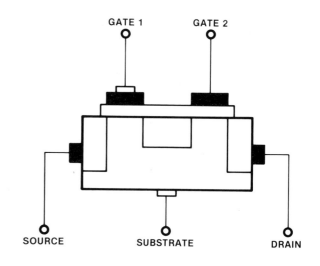

Figure 10-35. The current through a dual-gate MOSFET can be cut off by either gate.

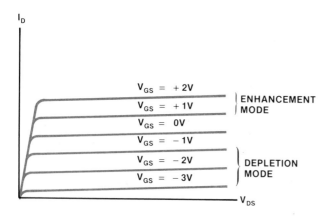

Figure 10-34. A typical set of characteristic curves for a depletion-enhancement MOSFET shows the operating ranges for the enhancement and depletion modes.

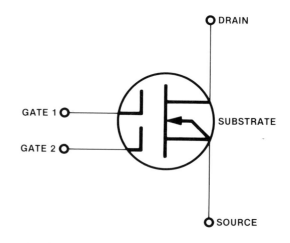

Figure 10-36. Shown above is the schematic symbol for a dual-gate MOSFET.

Amplifier Coupling

Coupling is the joining of two or more circuits so that power can be transferred from one amplifier stage to another. The three basic techniques for coupling amplifier stages together and to loads are *capacitive coupling, transformer coupling,* and *direct coupling.* No one coupling technique is always best because frequency, impedance-matching, cost, size, and weight must be considered when choosing a coupling technique for different applications.

Capacitive Coupling. Capacitive coupling has the disadvantage of poor impedance-matching. It also limits the lower frequency response of the amplifier. Despite these drawbacks, capacitive coupling is often used because it is inexpensive, and few components are required. If signal losses are encountered, they are generally offset by adding another transistor stage. Another reason why capacitive coupling is used is because it can block or isolate the bias circuits of each stage. The series coupling capacitor presents an open circuit to the flow of DC.

Figure 10-37 shows the effect a series coupling capacitor has in transferring a signal from a source to a load. When a 5-volt source is attached to this circuit, capacitor *C* charges to 5 volts. Current flows through *R* only during the charging of *C*, causing a voltage to appear momentarily across *R*. Once *C* is charged, the voltage across it will be 5 volts, and the voltage across *R* will be zero. If the source voltage is increased to 10 volts, *C* will increase its charge to 10 volts. The charging current will produce another momentary 5-volt pulse of voltage across *R*. See Figure 10-38. If source voltage is reduced to 3 volts, *C* will discharge to 3 volts. The discharge current will produce a momentary pulse of 7 volts across *R* because it is discharging to a lower voltage. See Figure 10-39.

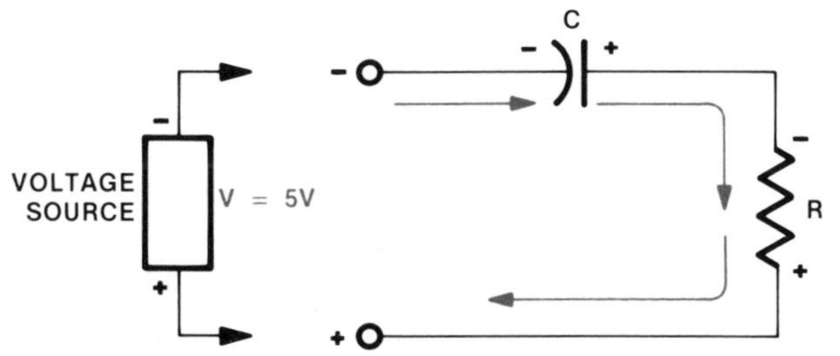

Both DC and AC voltage can be present in the circuit. See Figure 10-40. The input signal of 2 VAC varies around a DC level of 5 volts. The AC signal causes the input to vary between 7 and 3 volts.

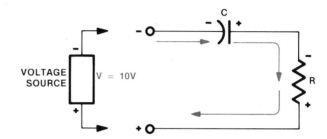

Figure 10-38. If the source voltage is increased to 10 volts, *C* will increase its charge to 10 volts. The charging current will produce another momentary 5-volt pulse across *R*.

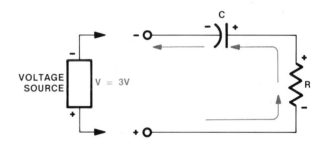

Figure 10-39. If the source voltage is reduced to 3 volts, *C* will discharge to 3 volts. The discharge current will produce a momentary pulse of 7 volts across *R*.

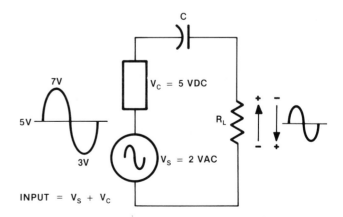

Figure 10-37. The 5-volt source attached to this circuit causes the capacitor *C* to charge to 5 volts. Current flows through *R* only during the charging time of *C*.

Figure 10-40. The circuit effectively blocks the 5-VDC source and allows the AC signal to pass on to load resistor R_L.

The capacitor charges and discharges according to the varying AC signal, assuming the capacitance value has been selected to pass the lowest frequency. Note that the output varies around zero level and the DC is effectively removed because the capacitor blocks the DC component. Because of this effect, the capacitor is called a blocking capacitor, or coupling capacitor. Blocking capacitors are used to connect one section of a multistage amplifier to another. See Figure 10-41.

Transformer Coupling. For maximum power transfer between amplifier stages, or between a stage and the load, correct impedance-matching is required. Special interstage transformers are used to match input and output impedances between stages. Transformers are also used to match load impedance to amplifier output impedance. Figure 10-42 shows the use of transformer coupling. Capacitor C_2, in this circuit, is required to prevent the fixed bias of transistor Q_1 from shooting through the transformer secondary T_1.

Transformer coupling can provide a high level of circuit efficiency. However, the weight, size, and cost of the transformers rule them out for many applications. Also, the frequency response is not as good as with capacitive coupling circuits, especially at higher frequencies.

Direct Coupling. In many industrial circuits, it is necessary to amplify very low frequency signals. Capacitive and transformer coupled amplifiers have poor frequency response at low frequencies because both of them block low frequency signals. Therefore, direct coupling must be used when low frequencies are involved. Direct coupling is also used where the DC value and the AC value of a signal must be retained. The directly coupled amplifier provides a frequency response that ranges from zero hertz (DC) to several thousand hertz.

Figure 10-43 shows a directly coupled amplifier. In this circuit, the collector of Q_1 is connected directly to the base of Q_2. The collector load resistor R_2 also acts as a bias resistor for Q_2. Any change of bias current is amplified by the directly coupled circuit, which is very sensitive to temperature changes. This disadvantage can be overcome with bias stabilization

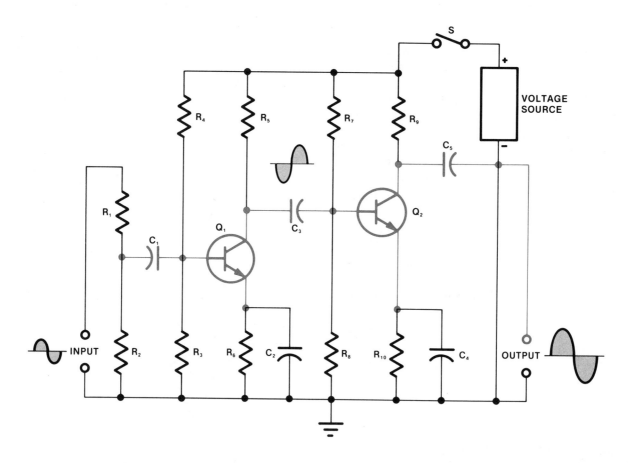

Figure 10-41. Blocking capacitors, or coupling capacitors, are used to keep DC voltage from passing from one stage to another, while the AC signal passes easily.

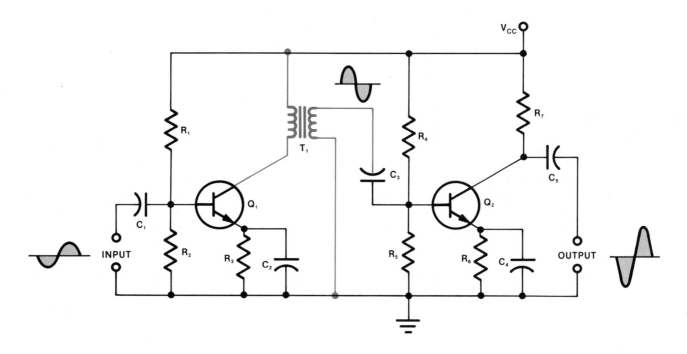

Figure 10-42. Transformer coupling provides a means of impedance-matching between stages.

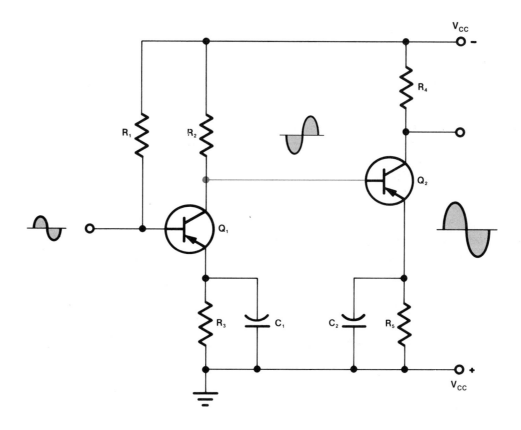

Figure 10-43. Direct coupling must be used when low frequencies are involved. It must also be used when the DC value and the AC value of a signal must be retained.

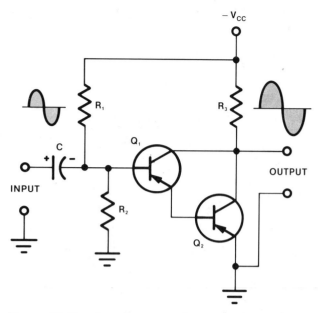

Figure 10-44. An advantage of the Darlington circuit is that it uses fewer components. Also, there is no loss incurred in the coupling.

circuits (see Chapter 6). Another disadvantage is that each stage may require different bias voltages for proper operation.

Darlington Circuit

It is often desirable to have a broad frequency range in a transistor amplifier. The Darlington circuit uses two bipolar transistors and direct connection without using any other type of coupling circuit. See Figure 10-44. The advantages of the Darlington circuit are extensive. Fewer components are required, and there is no loss incurred in the coupling. The Darlington circuit can be used as an output stage of a transistor amplifier that is driven by a stage of low or intermediate power. In a typical audio amplifier, coupling of this type of circuit requires very large capacitors. Therefore, it is inefficient. By using the Darlington circuit, smaller and lighter bipolar transistors may be used, and less power is dissipated.

Chapter 10 - Review Questions

1. What are the two categories of FETs?
2. What do the letters G, D, and S stand for in reference to FETs?
3. Name four uses of FETs.
4. What does "pinched off" mean?
5. What is the control element of a JFET?
6. What is the ohmic region of the JFET?
7. List the three basic configurations for JFETs.
8. What is the phase relationship between the input and the output of a common-source amplifier?
9. Where are common-gate amplifiers typically used?
10. Name three applications for JFETs.
11. What is a MOSFET?
12. What are the two MOSFET operation modes?
13. Explain the operation of an enhancement MOSFET.
14. Explain the operation of a depletion-enhancement MOSFET.
15. What precaution must be observed when installing or removing MOSFETs?
16. What is a multistage amplifier?
17. What are the major types of amplifier coupling?
18. What are the advantages of using a Darlington circuit?

11 INTEGRATED CIRCUIT (IC)

Integrated circuits (ICs) are popular because they provide a complete circuit function in one small semiconductor package. (ICs are often called chips, which are actually a component part of the IC.) See Figure 11-1. Although many processes have been developed to create these devices, the end result has always been a totally enclosed system with specific inputs and specific outputs.

Because of the nature of ICs, the technician must approach ICs in an entirely different manner from individual solid state components. An IC is a system within a system. The entire system of an IC, and what it does, must first be understood. Data books and manufacturer's specification sheets can usually provide this information. When these are not available, the inputs and outputs of the system must be studied by using meters and an oscilloscope. Troubleshooting ICs also requires knowledge of how the system functions, and what the input and output should be. Since ICs cannot be repaired, they must be replaced if they are defective.

Key Words

Active filter	Dual supply	Offset error
AND gate	Flat-pack	Open-loop
Astable multivibrator	High-pass filter	Operational amplifier (OP amp)
Bipolar supply	IC timer	OR gate
CAD/CAM	Integrated circuit (IC)	Oscillator
Closed-loop	Large-scale integration (LSI)	Planer technology
Comparator	Low-pass filter	Slew rate
Compensation	Mini-DIP	Summing amplifier
Current-to-voltage OP amp	Monostable multivibrator	Unity
Differential amplifier	NAND gate	Very large-scale integration (VLSI)
Digital logic	NOR gate	Voltage-to-current OP amp
DIP extractor	Nulling	555 IC
Dual-in-line package (DIP)		

ADVANTAGES AND DISADVANTAGES OF ICs

ICs have the following advantages:

1. In addition to their small size and light weight, they are extremely reliable.

2. They consume less power and cost less to operate than conventional components.

3. They generate less heat than conventional components and do not require as much cooling.

4. Because of their short lead lengths, signals may travel rapidly.

5. In large quantities, ICs are less expensive than conventional components.

Although ICs have many advantages, there are a few disadvantages:

1. They are low-power devices and are found mainly in the control circuit rather than in places where heavy currents are switched.

2. High voltages tend to break down ICs.

3. They cannot be repaired. Therefore, when an IC fails or is damaged, it must be replaced.

IC PACKAGES

Figure 11-2 shows various IC packaging. Their shapes and sizes range from standard transistor shapes, such as the TO-3 and TO-220 packages, to the latest in *large-scale integration (LSI)*. (Metal-oxide substrate, MOS, is a type of LSI.) When space is a premium, there are even ICs designed with *flat-pack* construction.

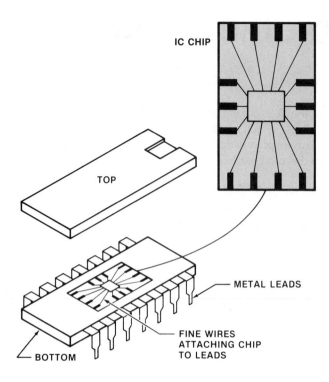

Figure 11-1. An integrated circuit (IC) provides a complete circuit function in one semiconductor package.

The *dual-in-line package (DIP)* with 14, 16, or 24 pins is the most widely used configuration. The *mini-DIP* is a smaller dual-in-line package with 8 pins. A modified TO-5 is available with 8, 10, or 12 pins.

The housings for ICs may be metal, plastic, or ceramic. Ceramic is used in applications where high temperatures may be a factor.

The appearance of these devices varies, and may give an initial impression of being one particular type of device. Therefore, it is essential that part numbers and manufacturer's literature be reviewed before working on the device.

PIN NUMBERING SYSTEM

All manufacturers use a standardized pin numbering system for their devices. When unsure about pin numbering patterns, consult the manufacturer's data sheets. Figure 11-3 shows the IC pin numbering sequence for the TO-220, TO-3, and TO-5 packages. For the TO-5 and modified TO-5, the leads are located in reference to the metal tab.

Dual-in-line packages and flat-packs have index marks and notches at the top for reference. Before removing an IC, note where the index mark is in relation to the board or socket to aid in installation of the unit. The numbering of the pins is always the same. The notch is at the top of the chip. To the left of the notch is a dot which is in line with pin 1. The pins are numbered counterclockwise around the chip. See Figure 11-4.

IC FABRICATION

A unique feature of the IC is that the components are processed simultaneously from common materials. A wide variety of fabrication techniques exist. However, the most widely used technique is based on silicon *planer technology* (process) developed for transistors. Planer technology is popular because it provides high quality semiconductor devices.

The basic steps of the planer process are shown in sequence in Figures 11-5 through 11-11. The base material is a uniform single crystal of N-type or P-type silicon. See Figure 11-5. Through a diffusion process, impurities can be added to the base material at specific depths and widths. Vertical depth penetration of the impurities into the base material is controlled by the diffusion temperature and time. The lateral, or width, control of the diffusion process is controlled by a combination of masking properties of silicon dioxide with photochemical techniques. (Masking on a semiconductor is similar to the masking tape applied to a car before it is painted. Only the areas left exposed

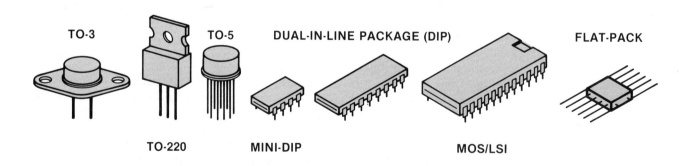

Figure 11-2. A variety of packaging is available for ICs.

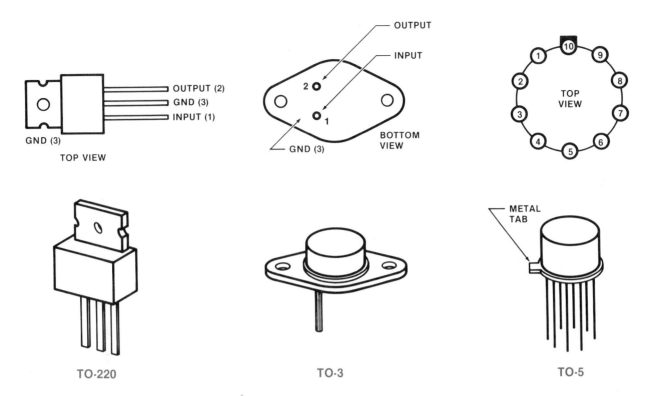

Figure 11-3. ICs are found in conventional TO-220, TO-3, and TO-5 packages. For the TO-5 and modified TO-5 packages, the leads are located in reference to the metal tab.

are ready for diffusion just as only the exposed areas on a car are ready for painting.)

When N-type materials are diffused into P-type base material, isolated circuit elements (diodes) are constructed. See Figure 11-6. The diodes formed in this manner accomplish electrical isolation between circuit elements. Diffusion of additional P-type and N-type regions can form a transistor. See Figure 11-7. Once formed, the silicon wafer is then coated with an insulating oxide layer, or mask, which can be opened specifically to permit metallization and interconnection. See Figure 11-8. If resistors are required, the N-type emitter diffusion is omitted and two ohmic contacts are made to a P-type region to form the resistor diffusion. See Figure 11-9. When capacitors are required, the oxide, or mask material, is used as a dielectric for the capacitor. See Figure 11-10. When the components are combined together, the device becomes an integrated circuit. See Figure 11-11. In this case, the complete chip contains a transistor, resistor, and capacitor.

TYPES OF IC SYSTEMS

An IC is a system made of a self-contained interconnection of several components. This system is designed to accomplish certain tasks. Each chip is designed for certain functions that must be accomplished by an electronic system. These include amplifiers, oscillators, timers, and many more. Since it is impossible to know the function of all the ICs available, the technician must use data books and manufacturer's specification sheets to determine the function of the IC. By understanding how a few common ICs operate, this knowledge should be used to obtain the information needed to understand most ICs.

Operational Amplifier (OP Amp)

The *operational amplifier (OP amp)* is one of the most widely used ICs. An OP amp is a very high-gain, directly coupled amplifier that uses external feedback to control response characteristics. An example of this feedback control is gain. The gain of an OP amp can be controlled externally by connecting feedback resistors between the output and input. A number of different amplifier applications can be achieved by selecting different feedback components and combinations. With the right component combinations, gains of 500,000 to 1,000,000 are common. OP amps are very versatile and can also be used as a voltage follower, inverter, differentiator, integrator, adder, subtractor, or phase shifter.

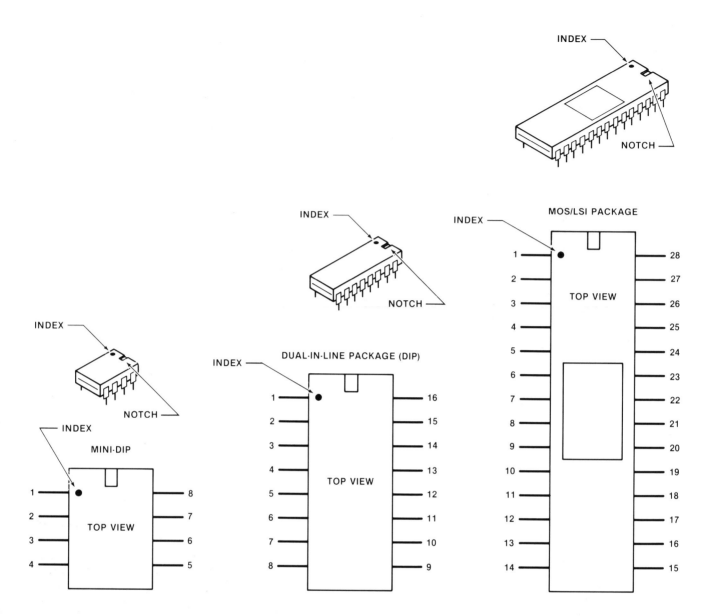

Figure 11-4. Dual-in-line packages (DIPs) and flat-packs have index marks and notches for reference. The numbering of the pins is always the same. The notch is at the top of the chip. To the left of this notch is a dot which is in line with pin 1. The pins are numbered counterclockwise around the chip.

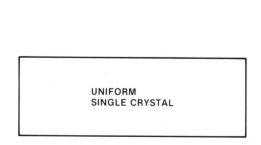

Figure 11-5. The base material for an IC is a uniform crystal of N-type or P-type silicon.

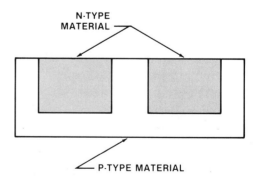

Figure 11-6. When N-type materials are diffused into P-type base material, isolated circuit elements are constructed. Here, two diodes have been formed in the P-type material.

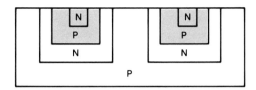

Figure 11-7. By diffusing additional P-type and N-type regions, transistors are formed.

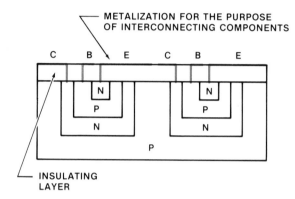

Figure 11-8. Once formed, the silicon wafer is then coated with an insulating oxide layer, which can be opened specifically to permit metallization and interconnection.

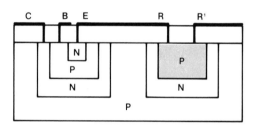

Figure 11-9. If resistors are required, the N-type emitter diffusion is omitted and two ohmic contacts are made to a P-type region to form the resistor.

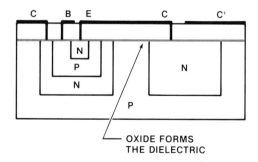

Figure 11-10. When capacitors are required, the oxide itself is used as a dielectric for the capacitor.

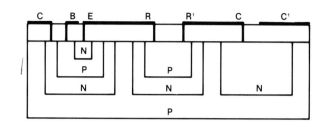

PICTORIAL

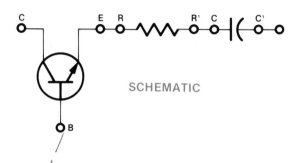

SCHEMATIC

Figure 11-11. When the components are combined together, the device becomes an IC. In this case, the complete chip contains a transistor, resistor, and capacitor.

OP Amp Schematic Symbols. The schematic symbol for an OP amp may be shown in two ways. See Figure 11-12. In each case, the two inputs of the OP amp are the inverting (−), and the non-inverting (+). The two inputs are usually drawn as shown, with the inverting input at the top. The exception to the inverting input being at the top is when it would complicate the schematic. In either case, the two inputs should be clearly identified by polarity symbols on the schematic symbol.

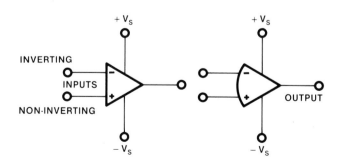

Figure 11-12. The schematic for an OP amp may be shown in two ways.

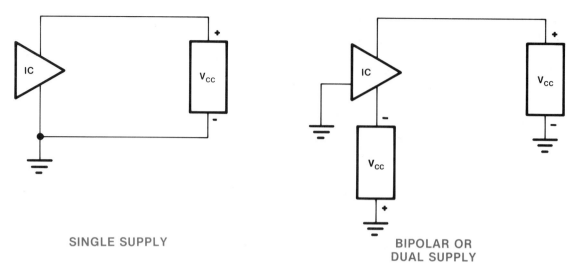

SINGLE SUPPLY   BIPOLAR OR
                                 DUAL SUPPLY

Figure 11-13. Voltage sources for OP amps can be single supply. However, they are usually bipolar or dual supply.

Voltage Sources for OP Amps. ICs need DC operating voltages just as other solid state devices do. The DC voltage pins on some ICs are labeled V_{CC}. Others have pins labeled Input Voltage (V_I) or Voltage Supply (V_S). DC voltage ratings in IC data books are labeled all three ways.

The voltage source for OP amps can be single supply. However, they are usually *bipolar* or *dual supply*. See Figure 11-13. The voltage supplies typically range from ±5 volts to ±15 volts. That is, one supply is +5 to +15 volts with respect to ground. The other

supply voltage is −5 to −15 volts with respect to ground. See Figure 11-14.

CAUTION: Never reverse the power supply polarity to an OP amp. Applying a negative voltage to the positive pin or a positive voltage to the negative pin, even momentarily, will result in destructive current flow through the OP amp.

No matter what type of power supply is used, most OP amp manufacturers suggest using bypass capacitors on the power supply leads. See Figure 11-15. The recommended capacitor size is about 0.1 farad (F).

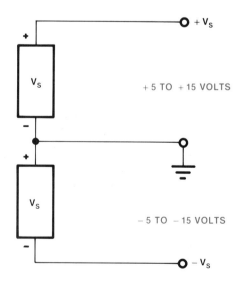

Figure 11-14. Voltage supplies typically range from ±5 volts to ±15 volts. One supply is +5 to +15 volts with respect to ground. The other supply is −5 to −15 volts with respect to ground.

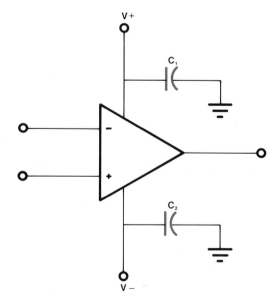

Figure 11-15. No matter what type of power supply is used, most OP amp manufacturers suggest using bypass capacitors on the power supply leads.

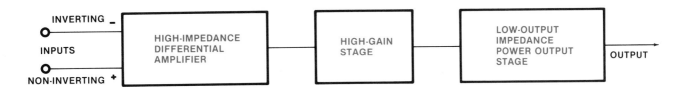

Figure 11-16. Internally, an OP amp consists of three major parts: a high-impedance differential amplifier, a high-gain stage, and a low-output impedance power-output stage.

It should be noted that in some schematic circuits, the IC power supply connection is not shown on each symbol. In such a case, it is shown somewhere on the schematic, and the technician should know that each IC must be powered from this source, unless otherwise stated.

Internal OP Amp Operation. Internally, an OP amp has three major parts. See Figure 11-16. It consists of a high-impedance *differential amplifier*, a high-gain stage, and a low-output impedance power-output stage. The differential amplifier provides the wide bandwidth and the high impedance. The high-gain stage boosts the signal. The power-output stage isolates the gain stage from the load and provides for the power output.

The operation of the differential amplifier is unique. See Figure 11-17. Currents to the emitter-coupled transistors Q_1 and Q_2 are supplied by the source Q_3. When manufactured, the characteristics of Q_1 and Q_2, along with their biasing resistors (R_1, R_2, and R_3), are closely matched to make them as equal as possible.

As long as the two input voltages, A and B, are either zero or equal in amplitude and polarity, the amplifier is balanced because the collector currents are equal. When balanced, zero voltage difference exists between the two collectors.

The sum of the emitter currents is always equal to the current supplied by Q_3. Thus, if the input to one transistor causes it to draw more current, the current in the other decreases and the voltage difference between the two collectors changes in a differential manner. See Figure 11-18. The differential swing, or output signal, will be greater than the simple variation that can be obtained from only one transistor. Each transistor amplifies in the opposite direction so that the total output signal is twice that of one transistor. This swing will then be amplified through the high-gain stage, and matched to the load through the power-output stage.

OP Amp Characteristics. The non-inverting input (+) of an OP amp is used in creating an *open-loop* voltage

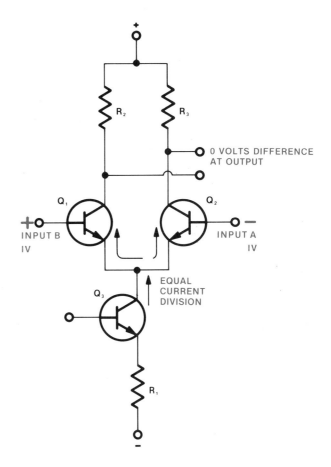

Figure 11-17. In this circuit, currents to the emitter-coupled transistors (Q_1 and Q_2) are supplied by the source Q_3. As long as the two input voltages, A and B, are either zero or equal in amplitude and polarity, the amplifier is balanced because the collector currents are equal. When balanced, a zero voltage difference exists between the two collectors.

follower (source follower) amplifier. The inverting input (−) of an OP amp is used in creating a *closed loop*, or feedback inverting, amplifier capable of an 180° phase shift.

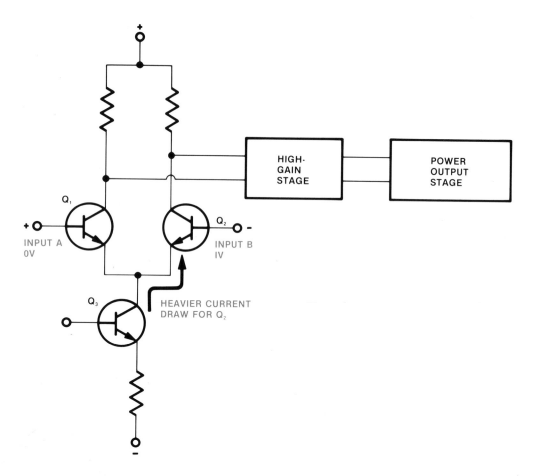

Figure 11-18. The sum of the emitter currents is always equal to the current supplied by Q_3. Thus, if the input to transistor Q_2 draws more current than Q_1, the current to Q_1 decreases and the voltage difference between the collectors of Q_1 and Q_2 changes in a differential manner.

Voltage Follower (Open-Loop). The voltage follower, or source follower, is a non-inverting amplifier. See Figure 11-19. The output voltage V_0 is an exact reproduction of the input voltage V_1.

The function of the voltage follower is identical to the source followers created by bipolar transistors and FETs. The circuit is used to impedance-match an input signal to its load. In the case of the voltage follower, the input impedance is high and the output impedance is low. Note that the voltage follower has no input or feedback components. Because no feedback components are used in this type of circuit, the amplifier is operating in an open-loop condition.

Feedback Inverting Amplifier (Closed-Loop). A feedback inverting amplifier produces a 180° phase inversion from input V_1 to output V_0. See Figure 11-20. When a positive-going voltage is applied to the input, a negative-going voltage will be produced at the output. The input signal is applied to the OP amp inverting input through R_1, while resistor R_2 serves as the feedback element.

The voltage gain of the feedback inverting amplifier can be less than, equal to, or greater than one (1.0). Its value depends on the values of resistors R_1 and R_2. Because a feedback component is used in this type of circuit, the amplifier is operating in a closed-loop condition. The closed-loop gain of the feedback inverting amplifier can be controlled by switching in different feedback resistors. See Figure 11-21. If resistors R_1 and R_F are equal, the OP amp can be used as a simple *unity* (no gain) signal inverter.

Compensation. Because high-gain OP amps usually use the feedback, the feedback must be controlled to ensure that the OP amp circuit is stable. If properly controlled, feedback should not affect changes in frequency nor cause oscillations if the input-output phase relationship changes.

If no phase *compensation* is furnished on an OP amp, the gain of the feedback signal may be greater than the input if the phase angle approaches 180°. In this case, feedback that is negative at low frequencies becomes positive at high frequencies, and unwanted

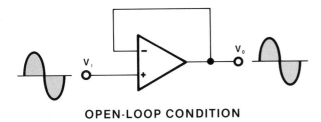

OPEN-LOOP CONDITION

Figure 11-19. The voltage follower, or source follower, has an output voltage V_0, which is an exact reproduction of the input voltage V_1. Because there are no feedback components in this circuit, the amplifier is operating in an open-loop condition.

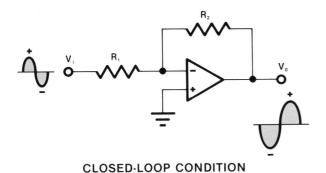

CLOSED-LOOP CONDITION

Figure 11-20. A feedback inverting amplifier produces a 180° phase inversion from input to output. Because a feedback component is used in this type of circuit, the amplifier is operating in a closed-loop condition.

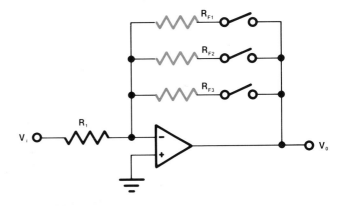

Figure 11-21. The closed-loop gain of the feedback inverting amplifier can be controlled by switching in different feedback resistors. If resistors R_1 and R_F are equal, the OP amp may be used as a simple unity (no gain) signal inverter.

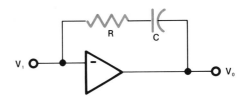

Figure 11-22. Compensation can be accomplished by a series resistor-capacitor combination. The amount of feedback in this circuit increases as the frequency increases because the reactance of the capacitor decreases. The upper limit for feedback is determined by the resistor value, which remains constant at high frequencies.

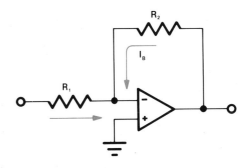

Figure 11-23. With proper bias, the feedback inverting amplifier has an input bias current through the input and feedback resistors, R_1 and R_2, with no signal applied. Current flow through these resistors produces a voltage drop which appears as a DC input voltage. The OP amp then amplifies this DC input voltage, compounding the problem.

oscillations may result.

To overcome this tendency toward unwanted oscillation, the frequency response and phase-shift characteristics of the OP amp must be compensated. Outside components, such as resistors and capacitors, are used for compensation. These components control the frequency response and phase-shift characteristics. See Figure 11-22. In this case, the amount of feedback increases as the frequency increases. This is because the reactance of the capacitor decreases. The upper limit for feedback is determined by the resistor value, which remains constant at high frequencies.

If an OP amp does not require compensation, it is identified as such in the manufacturer's specifications.

Offset Error/Nulling. Although extreme care is taken in fabricating an OP amp, a slight mismatch may still occur between the internal components. This is called *offset error*, and it creates a problem when using the OP amp in a DC circuit. The mismatch prevents the amplifier from having a zero output for a zero input.

Even with proper bias, the feedback inverting amplifier has an input bias current I_B through the input and feedback resistors, R_1 and R_2, with no signal applied. See Figure 11-23. The additional current flow

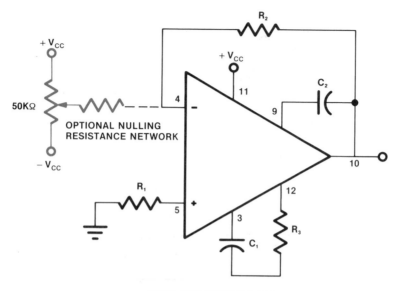

NOTE: THE NUMBERS AROUND THE
OP AMP REPRESENT ITS PIN NUMBERS

Figure 11-24. Nulling is a technique often used to correct offset error. Correction for offset error is made by using a nulling resistance network. With this network, the nulling potentiometer is adjusted for zero output with zero input.

through these resistors produces a voltage drop, which appears as a DC input voltage. The OP amp then amplifies this DC input voltage, compounding the offset error.

Nulling is a technique often used to correct offset error. Correction is made by using a nulling resistance network. See Figure 11-24. With this network, the nulling potentiometer is adjusted for zero output with zero input.

Bandwidth and Slew Rate. When a high frequency signal is fed into an OP amp, certain changes can take place with the output signal due to the high frequency. Some elements within the OP amp have capacitive characteristics. A certain time span is required to charge and discharge them. The capacitive effect prevents the output voltage from following the input signal immediately. Thus, these internal capacitances limit the rate at which the output voltage can change. This maximum time rate of change of the output is the *slew rate.* Its unit of measure is volts per microsecond.

Slew rate is the time that an amplifier takes to reach 90% of the steady-state output level for a given input. The slew rate of an OP amp is very fast. Slew rates in OP amps may vary from 0.1 volt per microsecond to 100 volts per microsecond. The slew rate of a feedback inverting amplifier depends on a number of factors, including the value of the closed-loop gain components.

Slew rate can limit bandwidth. The bandwidth is usually expressed at the highest frequency at which the amplifier develops its rated output without distortion.

Applications of OP Amps

OP amps are used for a variety of amplification applications, such as in audio amplifiers and video amplifiers. They are also ideal for a variety of industrial and commercial control systems.

By adding various external components, OP amp characteristics change to meet different circuit conditions. The number of circuit possibilities for OP amps appear limitless.

OP Amp as a Current-to-Voltage Converter. The *current-to-voltage OP amp* shown in Figure 11-25 makes use of the current sensitivity of the OP amp to measure very small currents. As shown, the circuit can provide 1 volt at the output for 1 microampere at the input.

This basic current-to-voltage converter is essentially an inverting amplifier without an input resistor. The input current is applied directly to the inverting input of the OP amp.

A practical example of a current-to-voltage converter is shown in Figure 11-26. Here, a thermistor varies the amount of current I_{IN} entering the inverting input of the OP amp. As the temperature increases, the resistance of the thermistor decreases, current to the input of the OP amp increases, and voltage at the output increases. As the temperature decreases, the resistance increases, current decreases, and the voltage output decreases.

Since the output of the circuit is now voltage, the OP amp voltage can be used to drive a MOSFET, which in turn, serves as the output switching element to the load. See Figure 11-27. Thus, a very small current source can be used through an OP amp and MOSFET as a very effective control device.

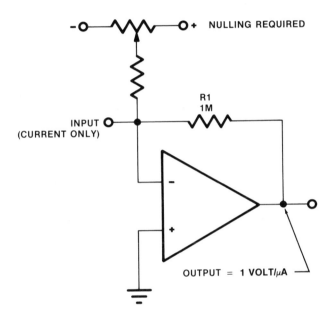

Figure 11-25. A current-to-voltage OP amp makes use of the current sensitivity of the OP amp to measure very small currents. A current-to-voltage converter is essentially an inverting amplifier without an input resistor.

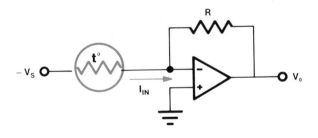

Figure 11-26. As the ambient temperature around the thermistor changes, the resistance of the thermistor varies. With a change in resistance, the input current also changes. An increased resistance means a reduction in current, while a decreased resistance means an increase in current. These current changes are turned into voltage changes at the output of the current-to-voltage OP amp circuit.

OP Amp as a Voltage-to-Current Converter. In certain circuits, a change in voltage becomes the reference for change in a circuit. When this is the case, a voltage-to-current OP amp is used. See Figure 11-28. The output voltage of the bridge circuit is a function of the degree of imbalance present in the input bridge. For the bridge circuit, an imbalance can be created by changing the pressure on the pressure sensor (transducer). The rheostat determines the pressure-set limits.

The unit can operate directly from the AC supply since it incorporates a step-down transformer and single-phase rectifier. When the pressure drops, the resistance of the pressure sensor decreases. Terminal 3 of the OP amp becomes more positive than terminal 2. Under this condition, the output current at pin 6 causes the triac to conduct. With the triac conducting, power is applied to coil M1, which turns ON the air compressor. When the pressure in the tank is brought up to the preset limit, the value of the pressure sensor increases, balancing the bridge, and the air compressor is shut OFF.

Differential OP Amp for Thermostat Control. Figure 11-29 shows an OP amp used as a differential amplifier. This particular OP amp is an example of the ones used in differential-thermostat controls for solar heating units. In this circuit, small signal changes can be detected easily, and are amplified for signaling purposes.

When an OP amp has signals of equal amplitude and polarity applied to each of its inputs simultaneously, the output will be zero. In the case of the

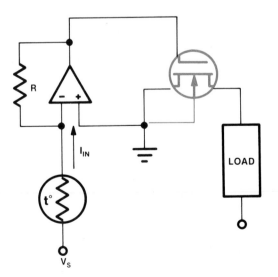

Figure 11-27. Since the output of the current-to-voltage OP amp is voltage, the thermistor and OP amp can be used with a MOSFET to form an effective switching element to a load.

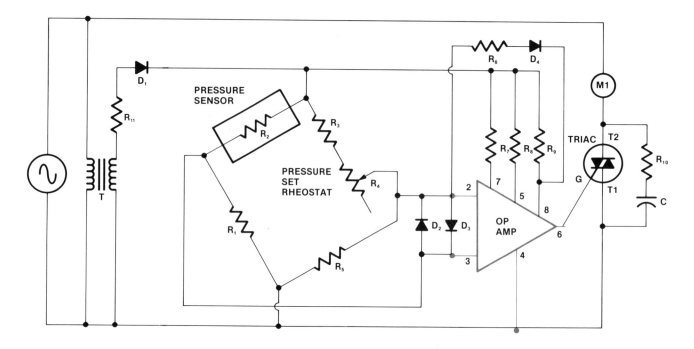

AIR COMPRESSOR ELECTRICAL CIRCUIT

Figure 11-28. The bridge circuit provides the voltage difference to make this air compressor circuit operate. When the pressure on the sensor is too low, the voltage-to-current OP amp triggers the triac, which powers coil M1, turning ON the air compressor. As the pressure reaches the preset limit determined by the rheostat, the bridge is balanced. With the bridge balanced, there is no voltage input and no current output. This shuts OFF the triac, which controls the coil, and shuts OFF the air compressor.

circuit in Figure 11-29, the bridge must be balanced in order for the OP amp output to be zero. The bridge is formed by the tank sensor, collector sensor, and resistors R_1, R_2, and R_3.

If the tank senses a reduction in heat, the bridge will become unbalanced, with the difference being presented to the input of the OP amp. The signal is amplified to the output. The output is then connected to a relay that starts the pump. When enough hot water passes from the collector, past the tank, and heats the tank water, the bridge is again balanced and the relay is turned OFF.

OP Amp as an Integrator. Figure 11-30 shows an OP amp used as an *integrator*. It can integrate a voltage that varies with time. In control systems, it is often necessary to integrate or determine the total energy content of a series of signal pulses during some fixed period of time. An electronic tachometer, for example, converts the rotational speed of a shaft into a series

of pulses of fixed amplitude and duration. The faster the speed of rotation is, the more pulses there will be. Each of these pulses contributes to the charge across capacitor C (connected to the OP amp) during this time period. The output voltage of the OP amp will be directly proportional to the number of pulses occurring during that time. At the end of each time period, the electronic reset switch in the OP amp is closed momentarily to discharge the capacitor.

OP Amp as a DC Voltmeter. Figure 11-31 shows a solid state DC voltmeter circuit. If the 5K output resistor is changed to a different value, the sensitivity of the basic circuit can be changed. It is not unusual to have a full-scale range of 0.1 volt to 100 volts, with 100,000 ohms per volt input sensitivity.

The circuit has a null balance circuit since the typical DC offset multiplied by a gain of 100 would produce a significant zero offset on the meter. In this case, with the input shorted, the null balance potentiometer is

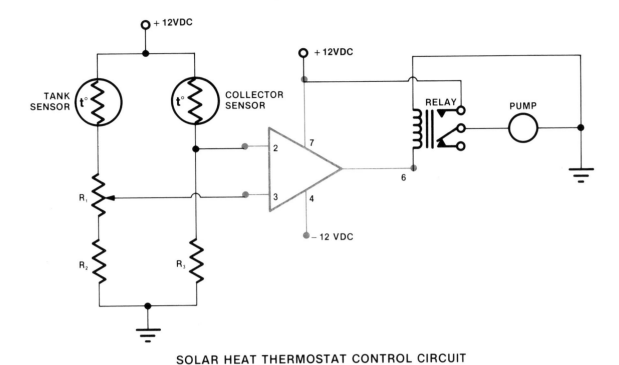

SOLAR HEAT THERMOSTAT CONTROL CIRCUIT

Figure 11-29. When the bridge is unbalanced, the OP amp detects the difference and amplifies it. The signal is applied to the relay, which in turn controls the pump. As the bridge is balanced, the relay turns OFF the pump. The temperature of the water in the tank and collectors determines whether the bridge is balanced or unbalanced.

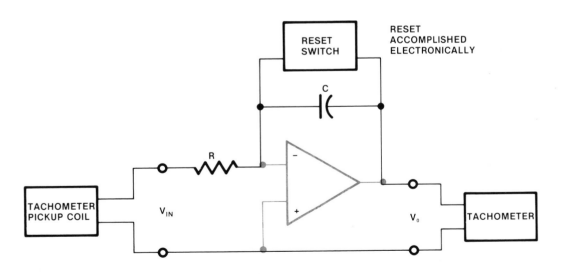

Figure 11-30. The OP amp as an integrator drives an electronic tachometer. The integrator takes the series of pulses coming from the tachometer pickup coil, and produces an output voltage that is directly proportional to the number of pulses occurring during that time. Capacitor C stores the pulse. The electronic reset discharges C.

adjusted to obtain a zero reading on the meter. This particular DC voltmeter is ideal for solid state testing since it has the necessary low-voltage scale readings and high impedance required for solid state devices.

OP Amp Summing Amplifier. A *summing amplifier* can be created by connecting several parallel input resistors to the inverting input of an OP amp. See Figure 11-32. If resistors R_1, R_2, and R_3 are equal to the feedback resistance R_F, the output voltage V_0 will be equal to the combined voltage drop across R_1, R_2, and R_3. Mathematically, the voltage output V_0 is:

$$V_0 = -(V_1 + V_2 + V_3)$$

NOTE: The voltage output is negative since the input was applied to the inverting input.

If gain is required from the circuit, the value of R_F must be larger than the resistance of R_1, R_2, and R_3.

OP Amp Sine-Wave Oscillator. An OP amp may be used to create an *oscillator* (see Oscillators). One of the more popular types is a Wien-bridge oscillator. See Figure 11-33. The resistor-capacitor combinations R_1-C_1 and R_2-C_2 provide a positive feedback path around the OP amp. The resistor-lamp combination R_3-L_1 provides a negative feedback to the OP amp. The positive feedback causes the circuit to oscillate with a sine-wave output. The negative feedback helps to regulate and stabilize the amplitude of the sine-wave output.

The frequency of the Wien-bridge oscillator can be determined mathematically, or by using a measuring instrument called a frequency counter.

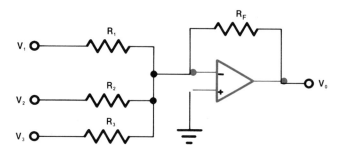

SUMMING AMPLIFIER CIRCUIT

Figure 11-32. In a summing amplifier, the output is equal to the sum of voltages across the parallel input resistors.

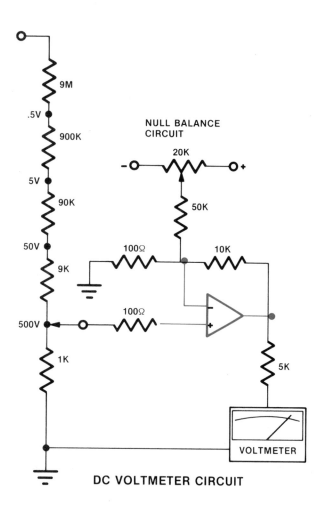

DC VOLTMETER CIRCUIT

Figure 11-31. When an OP amp is used in a DC voltmeter circuit, a full scale range of 0.1 volt to 100 volts, with 100,000 ohms per volt sensitivity, is possible.

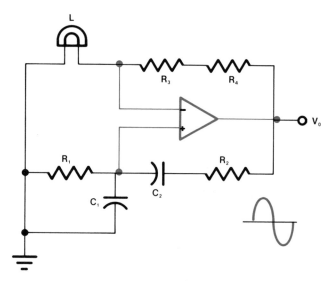

WIEN-BRIDGE (SINE WAVE) OSCILLATOR CIRCUIT

Figure 11-33. With positive feedback, an OP amp can be used to create a sine-wave oscillator.

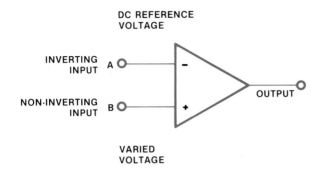

Figure 11-34. The comparator circuit is used to compare an input voltage with a DC reference voltage.

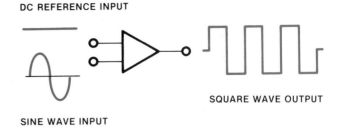

Figure 11-36. A sine wave to square wave converter is created by placing a varying sine wave on the non-inverting input, and a fixed DC reference on the inverting input.

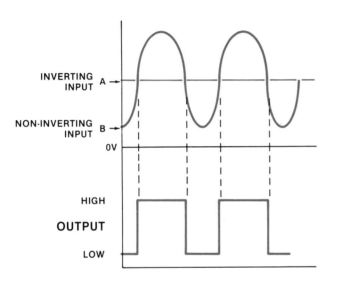

Figure 11-35. Whenever the signal at the inverting terminal (A) is more positive than the non-inverting terminal (B), the output of the OP amp will be a low voltage. Whenever the input signal at B is more positive than at A, the output of the OP amp will be a high voltage.

OP Amp as a Comparator. The *comparator* circuit is used to compare a varied input voltage with a DC reference voltage. See Figure 11-34. The final output of the comparator will change. The change depends upon whether the varied input signal is above or below the DC reference voltage to which it is compared.

Whenever the signal at the inverting terminal (A) is more positive than the non-inverting terminal (B), the output of the OP amp will be a low voltage. Whenever the input signal at B is more positive than at A, the output of the OP amp will be a high voltage. See Figure 11-35.

An example of this OP amp operation is shown in Figure 11-36. The voltage on the inverting input is a

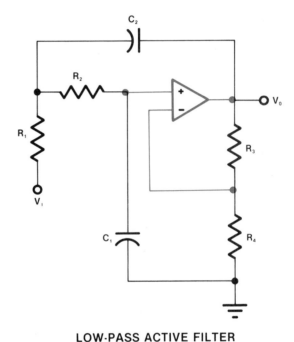

LOW-PASS ACTIVE FILTER

Figure 11-37. An active filter allows currents at certain frequencies to pass, while preventing the passage of others. The external components and types of OP amps determine if the circuit will be a high-pass or low-pass type.

DC voltage. The voltage on the non-inverting terminal is an AC sine wave, which varies above and below the DC value. The output is a square wave. This circuit is considered a sine wave to square wave converter.

OP Amp as an Active Filter. When an OP amp is used as an *active filter*, it allows currents at certain frequencies to pass through it. Meanwhile, it prevents the passage of others. With the proper network of resistors and capacitors, the active filter can block or pass a specific range of frequencies. The circuit of Figure 11-37 is a low-pass active filter. It is often called

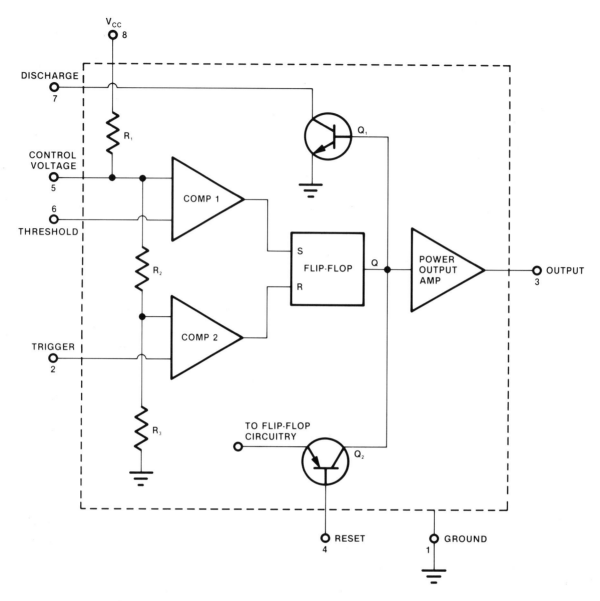

Figure 11-38. The 555 timer consists of a voltage divider network (R_1, R_2, and R_3), two comparators (Comp 1 and Comp 2), a flip-flop, two control transistors (Q_1 and Q_2), and a power-output amplifier.

a Sallen and Key Filter. A *low-pass filter* allows frequencies under a certain frequency to pass while blocking those above that frequency. A *high-pass filter* blocks low frequency signals and only passes high frequency signals.

IC Timers

An *IC timer* is a solid state timing device on a single chip. Two types of IC timers are the self-repeating timer and the externally triggered timer. A self-repeating timer is called an astable timer or *astable multivibrator*. An externally triggered timer is called a monostable timer or *monostable multivibrator*. (The

monostable timer is also called a *one-shot timer* or *one-shot multivibrator*.)

The *555 IC* is one of the most popular chips used for timing purposes. The 555 IC is a very stable IC that can operate either as an accurate monostable, or astable multivibrator. The 555 timer can be connected to time ON and OFF repeatedly by itself. It can also "time out" when the circuit receives an outside trigger signal. For the 555 IC to be fully functional, certain external components must be connected to the IC. The actual timing is done with an RC time-delay network.

The time it takes for the capacitor to charge or discharge is determined by the value of the resistance and the size of the capacitor. The accuracy of the timing sequence depends on the tolerance of the resistor and the leakage in the capacitor. Timing circuits, which are not critical, may vary as much as $\pm 20\%$ of the stated delay. Other precision timers may vary $\pm 1\%$ or less of the stated delay.

555 IC Operation. The 555 timer consists of a voltage-divider network (R_1, R_2, and R_3), two comparators (Comp 1 and Comp 2), a flip-flop, two control transistors (Q_1 and Q_2), and a power output amplifier. See Figure 11-38.

NOTE: A flip-flop is the electronic equivalent of a toggle switch. It has two outputs—one high and one low. When one is high, the other is low, and vice versa.

The comparators compare the input voltages to internal reference voltages that are created by the voltage divider, which consists of R_1, R_2, and R_3. Since the resistors are of equal value, the reference voltage provided by two resistors is two-thirds of the supply voltage (V_{CC}). The other resistor provides one-third of V_{CC}. The value of V_{CC} may change 9 volts, 12 volts, 15 volts, and so on, from chip to chip. However, the two-thirds/one-third ratio always remains the same. When the input voltage to either one of the comparators is higher than the reference voltage, the comparator goes into saturation and produces a signal that will trigger the flip-flop. In this IC circuit, the flip-flop has two inputs: S and R.

NOTE: The two comparators are feeding signals into the flip-flop. Comparator 1 is called the threshold comparator and comparator 2 is called the trigger comparator. Comparator 1 is connected to the S input of the flip-flop and comparator 2 is connected to the R input of the flip-flop.

Whenever the voltage at S is positive and the voltage at R is zero, the output of the flip-flop is high. Whenever the voltage at S is zero and the voltage at R is positive, the output of the flip-flop is low. The output from the flip-flop at point Q is then applied to transistors Q_1 and Q_2 and to the output amplifier simultaneously. If the signal is high, Q_1 will turn ON such that pin 7 (the discharge pin) will be grounded through the emitter-collector circuit. Q_1 will then be in a position to turn ON pin 7 to ground through the emitter-collector circuit.

NOTE: Pin 7 is called the discharge pin because it is connected to the timing capacitor. When Q_1 conducts, pin 7 is grounded and the capacitor can be discharged. See Figure 11-39.

Referring back to Figure 11-38, the flip-flop signal is also applied to Q_2. A signal to pin 4 can be used to reset the flip-flop. Pin 4 can be activated when a low-level voltage signal is applied. Once applied, this signal will override the output signal from the flip-flop. The reset pin (pin 4) will force the output of the flip-flop to go low, no matter what state the other inputs to the flip-flop are in.

The flip-flop signal is also applied to the power output amplifier. The power output amplifier boosts the signal and the 555 timer delivers up to 200 milliamps of current when operated at 15 volts. The output can be used to drive other transistor circuits, and even a small audio speaker. The output of the power output will always be an inverted signal compared to the input. If the input to the power-output amplifier is high, the output will be low. If the input is low, the output will be high.

555 IC Used as a Monostable Multivibrator (One-Shot Timer). Timing signals are needed to produce START/STOP commands, to synchronize pulses, and to produce special sequences. These signals are often produced with time-delay circuits, which include the 555 IC.

Some time-delay circuits produce an output until a certain time after the input appears. Monostable multivibrators can be used for this type of time-delay sequence. The monostable multivibrator has one stable state and one unstable state. An RC time constant determines the length of time that the circuit exists in the unstable state. The monostable multivibrator produces an output as soon as an input pulse appears. The duration of the output is determined by the time duration of the unstable state.

Figure 11-39 shows the external circuitry needed to operate the 555 IC as a one-shot timer. Two external components, variable resistance R and capacitor C, determine the amount of time delay. Capacitor C is usually held in the discharge state by transistor Q_1, which shorts C to ground. The timing cycle begins when a negative pulse is applied to the trigger input, pin 2. This negative pulse forces the flip-flop to go low on the output, which in turn removes the base bias to the discharge transistor Q_1. With Q_1 OFF, the short circuit (ground) is removed from capacitor C. Therefore, C is allowed to start charging with a time constant established by the values of R and C.

When the voltage across C reaches two-thirds of V_{CC}, as determined by the voltage divider, Comp 1 resets the flip-flop, returning it to its original high state. This change causes Q_1 to turn ON, discharging the external capacitor. With the capacitor discharging, the charging cycle stops. The resetting of the flip-flop drives the power-output amplifier, causing the output to return to its normal low operating condition.

The overall timing sequence is best shown by comparing input and output waveforms to the charging rate of the RC network. See Figure 11-40. It is seen here that a very short input pulse at pin 2 produces a relatively long output pulse at pin 3.

NOTE: The leading edge of the input pulse starts with the leading edge of the output pulse. The width

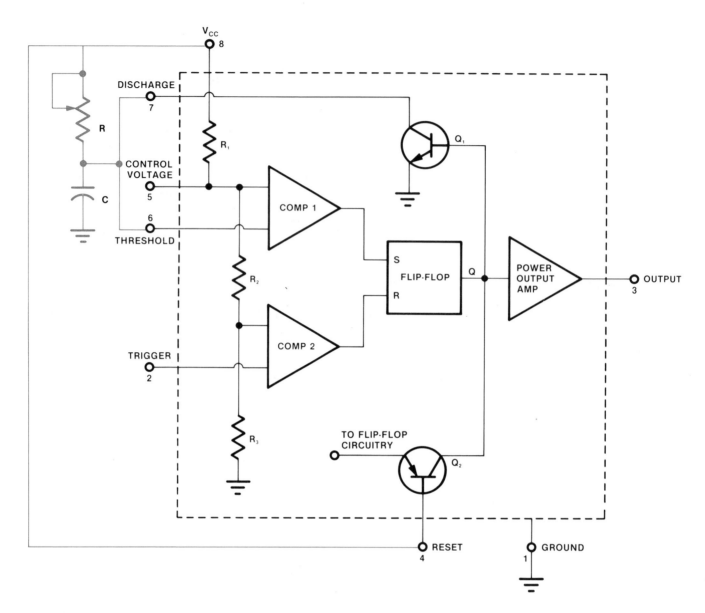

Figure 11-39. The external circuitry of resistor *R* and capacitor *C* determines the amount of time delay for the 555 IC timer.

of the output pulse is determined by the voltage produced across the capacitor, which is determined by the RC time constant.

555 IC Used in a Furnace Blower Control Circuit. The initial triggering signal comes from the thermostat. When the thermostat calls for cooling, the 555 IC circuit receives a signal. After an appropriate delay (for example, 4 minutes), the timer energizes a control relay. The control relay in turn, energizes the compressor contactor. If the trigger signal is removed from the circuit for any reason (power failure or thermostat OFF), the circuit will go through another 4-minute time cycle as the new signal appears.

Some furnace blowers use the time-delay feature to keep the blower ON for a certain amount of time after the resistance heater or gas flame has gone OFF. This type of circuit is usually activated by removing the trigger signal or by activating the reset pin on the IC timer.

555 IC Used in a Burglar Alarm Circuit. Electronic alarms are among the most widely used electronic devices. Almost all businesses use a burglar alarm. Many automobiles also have burglar alarms. Figure 11-41 shows a typical burglar alarm circuit. Smoke detectors are used in homes, apartment buildings, and businesses. Computers are equipped with fault alarm

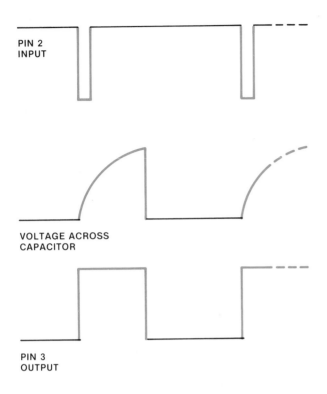

PIN 2
INPUT

VOLTAGE ACROSS
CAPACITOR

PIN 3
OUTPUT

Figure 11-40. The overall timing sequence of the 555 IC is shown by comparing input and output waveforms to the charging of the *RC* network.

systems used to prevent loss of valuable data.

Photoconductive cell R_1 and resistor R_2 of Figure 11-41 form a voltage divider. This divider produces the biases for the base-to-emitter junction of Q_1. With R_1 in darkness, its resistance will be high. Therefore, when current flows through the voltage-divider network, a large voltage will be dropped across the resistance developed by R_1. This voltage forward biases Q_1 since its base is now positive. With forward bias, Q_1 conducts and drops the voltage at trigger pin 2 to a low value. When pin 2 is low, the output of the flip-flop will be low and the output of the power amplifier will be high (approximately $+10$ volts). This is because the power amplifier not only amplifies, but also inverts $180°$, making the low input a high output. Since the voltage across pin 3 is approximately the same as V_{CC}, there is 0 volts across the relay K1. Thus, the relay is de-energized, and it prevents the burglar alarm signal from being connected to the speaker.

When light strikes R_1, its resistance drops to a much lower value. Reduced resistance causes a reverse bias on Q_1 by dropping the voltage at its base to a low value (less than 0.6 volts). With reverse bias applied to the base-emitter junction, Q_1 turns OFF. This raises the voltage at pin 2 to $+10$ volts, the supply voltage. With

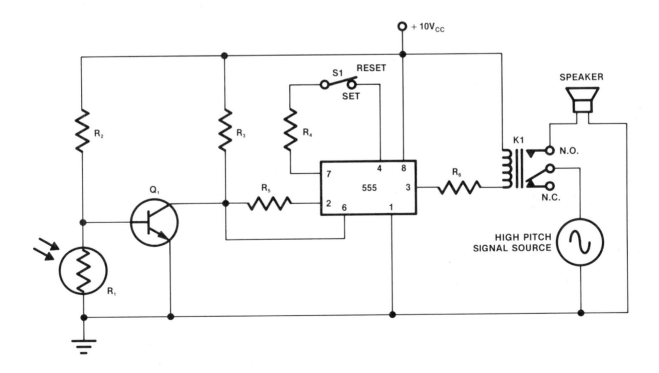

Figure 11-41. The 555 IC can be used as part of a burglar alarm circuit.

pin 2 high, the 555 output at pin 3 will be low. With pin 3 low, +10 volts will appear across the relay coil K1, and the relay will energize. This allows the burglar alarm signal to be applied to the speaker through the N.O. contacts of the relay K1. Thus, the burglar alarm will be activated. To shut OFF the alarm, the reset button must be pushed (closed). By closing the reset button, a pulse is applied to the reset of the 555 IC, putting the circuitry back to its original state.

Oscillators

Multivibrators are basic electronic circuits found in many types of equipment. Astable multivibrators are used extensively as waveform generators or oscillators. Astable multivibrator outputs are often used to control audible signals, such as sirens and other warning devices. Emergency vehicles often have sirens that are controlled electronically. Electronic circuits can change the pitch, cause the siren to warble fast or slow, or produce intermittent siren bursts.

In other applications, astable multivibrator square waves are used to control pulses of light. This is possible because square wave voltages act like electronic switches. The waves have straight sides that turn lights ON and OFF very quickly. This effect can be seen in the barricade flashers used on highways. Other examples are automobile 4-way flashers, digital-display scoreboards, and strobe lights used in the entertainment business.

Astable Multivibrator Operation. For the 555 IC to be used as an astable multivibrator, the chip must be continuously retriggered. The easiest way to retrigger the chip is to connect trigger input (pin 2) to the threshold input (pin 6). See Figure 11-42. The timing resistor R_1 and R_2 are added with their junction point connected to the discharge terminal (pin 7).

Before power is applied to this circuit, the trigger and threshold inputs are both below one-third of V_{CC}, the timing capacitor C is discharged, and the output is high. At the moment power is applied, C will begin charging.

C continues charging until its voltage reaches two-thirds of V_{CC}. At that point, comparator 1 (in the 555 IC) triggers the internal flip-flop, causing the capacitor to discharge through resistor R_2.

The time it takes to charge and discharge the capacitor is determined by the value of internal resistors, R_1, R_2, and R_3 of the IC. The total time required to complete one charge/discharge cycle is found by using the formula:

$$\text{(time) } t = 0.693 \ (R_1 + 2R_2)C$$

The frequency of oscillation is found by using the formula:

$$\text{(frequency) } f = 1.44/(R_1 + 2R_2)C$$

555 IC Used as a Thickness Monitor. The 555 in Figure 11-43 can use a light-controlled resistance element (photocell) to vary the frequency. Circuits like this can be used to monitor the thickness of semi-transparent materials on a production line. The material passes between the light source and the photocell. A change of thickness in the material causes a change in output frequency. This variation in frequency is then monitored by other electronic circuits. Finally, an indicator that notifies the operator of a problem in material thickness is activated.

555 IC Used as a Gas Alarm. Gas alarms are used, in addition to smoke detectors, to provide early warning of toxic gases and fumes from a fire or gas leak. The operation of the gas alarm circuit is dependent upon a semiconductor sensor whose electrical resistance changes when its active surface is exposed to gases, such as carbon monoxide, methane, and butane. The sensor element is normally enclosed in a small capsule and protected by a stainless-steel mesh. A low-power heater within the device activates the sensor element and purifies the element after exposure to gas.

A change in resistance is the key to triggering the alarm circuit using the 555 IC. See Figure 11-44. By

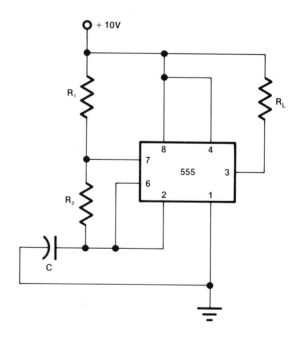

Figure 11-42. The circuit of a 555 IC can be used as an astable multivibrator.

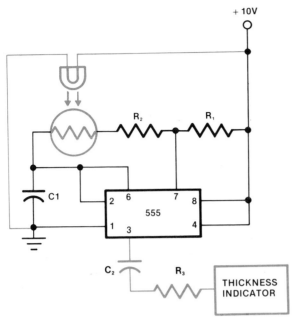

Figure 11-43. The 555 astable multivibrator can be used as a thickness monitor.

connecting the resistive element of the gas detector TGS between the positive supply and pins 2 and 6 of the 555 IC (No. 1), the 555 IC can be used as a comparator.

When the resistance of the TGS resistive element drops due to the presence of gas, the resistance ratio of the TGS element to R_1 and R_2 changes. Thus, the voltage at pins 2 and 6 is raised. The resistance of R_2 is adjusted so that voltage increase exceeds two-thirds V_{CC}. When that happens, the output of the No. 1 555 IC goes low, activating No. 2 555 IC. No. 2 555 IC is designed to operate as an astable alarm. It will drive an audible tone to the speaker, indicating the presence of gas.

When no gas is present, the resistance of the TGS element will be high enough so that the voltage applied by the diode will be below one-third V_{CC}. The output of No. 1 555 IC is high, therefore disabling the astable.

Normally, the sensor heater element receives its power from a stable voltage regulator at about 5 volts. To clean the element, however, a higher voltage is required. A clean/run switch is installed to allow the voltage regulator to be bypassed and a higher voltage applied for cleaning.

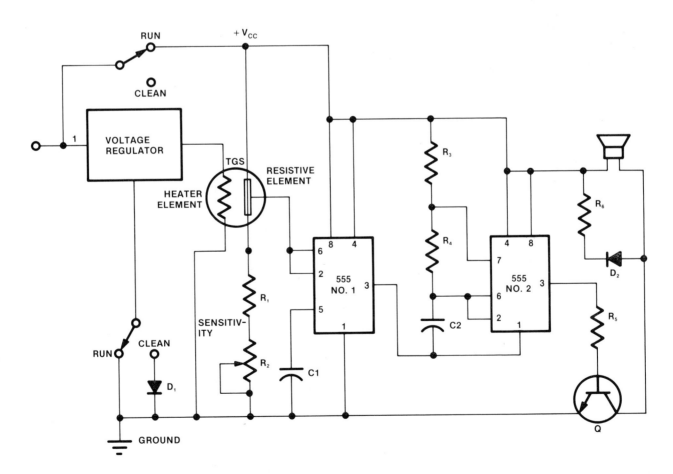

Figure 11-44. The 555 IC can be used as part of a gas alarm circuit.

Digital ICs

The two basic types of electronic signals are the analog signal and the digital signal. The analog signal, in the form of AC and DC voltages and currents, has been discussed throughout the text. These voltages and currents vary smoothly or continuously. A digital signal, on the other hand, is a series of pulses that changes levels between either the OFF state or ON state.

The analog and digital processes can be seen in a simple comparison between the light dimmer and light switch. A light dimmer varies the intensity of light from fully OFF to fully ON over a range of brightnesses. This is an example of an analog process. The standard light switch, on the other hand, has only two positions. It is either fully OFF or fully ON. This is an example of digital process. Electronic circuits that process these quickly changing pulses are digital or logic circuits.

When a technician becomes involved with a digital IC, the nature of electronics changes dramatically. The technician is leaving the "analog world" and is entering a "new world" controlled by *digital logic*. This section will not give the technician a complete understanding of digital electronics. Instead, it will give an orientation concerning the kinds of chips that are found in digital electronics. The four most common gates used in digital electronics are the *AND gate, OR gate, NAND gate,* and *NOR gate.*

AND Gate. The quad AND gate is one type of IC chip. See Figure 11-45. The manufacturer places four AND gates in one package—hence, the quad AND gate. By using the numbering system on the chip, any one, or all four AND gates may be utilized. In this case, voltage is applied to the circuit at pins 14 and 7.

The AND gate is a device with an output that is high only when both of its inputs are high. See Figure 11-46. To connect to the AND gate, pins 1, 2, and 3 of the quad AND gate chip could be used. Pins 1 and 2 are the input and pin 3 is the output. See Figure 11-47.

A practical application of an AND gate is in an elevator control circuit. See Figure 11-48. The elevator cannot move unless the inner and outer doors are closed. Once both doors are closed, the output of the AND gate could be fed to an OP amp, which in turn, fires a triac that starts the elevator motor. The electrical equivalent of an AND gate is shown in Figure 11-49.

OR Gate. Figure 11-50 shows a quad OR gate. The OR gate is a device with an output that is high if either or both inputs are high.

A practical application of an OR gate is in a burglar alarm circuit. If the front door or the back door is

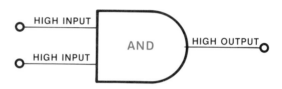

Figure 11-46. The AND gate is a device with an output that is high only when both of its inputs are high.

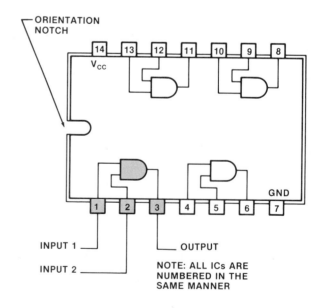

Figure 11-47. Pins 1, 2, and 3 on the quad AND gate chip could be used to make connections to the AND gate.

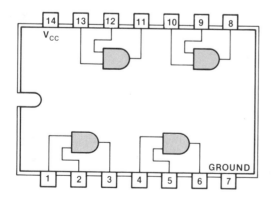

Figure 11-45. The quad AND gate allows the manufacturer to place four AND gates in one package.

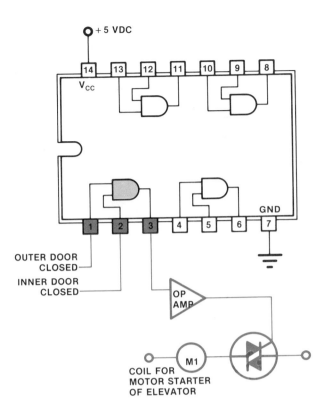

Figure 11-48. A practical application of an AND gate is in an elevator control circuit.

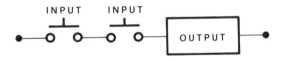

Figure 11-49. The electrical equivalent of an AND gate is two electrical push buttons in series.

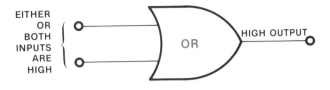

Figure 11-50. The OR gate is a device with an output that is high if either or both inputs are high.

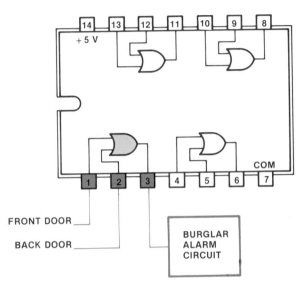

Figure 11-51. A practical application of an OR gate is in a burglar alarm circuit.

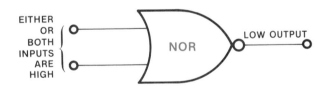

Figure 11-52. The NOR (NOT-OR) gate is the same as an inverted OR function. Thus, the NOR gate provides a low output if either or both inputs are high.

open, a signal is sent to the burglar alarm circuit. See Figure 11-51. The electrical equivalent of an OR gate is two push buttons in parallel.

NOR Gate. The NOR (NOT-OR) gate is the same as an inverted OR function. Thus, the NOR gate provides a low output if either or both inputs are high. See Figure 11-52. The NOR gate is represented by the OR gate symbol followed by a small circle indicating an inversion of the output. See Figure 11-53. The NOR gate is a universal building block of digital logic. It is usually used in conjunction with other elements to implement more complex logic functions. NOR gates are also available in quad IC packaging.

NAND Gate. The NAND (NOT-AND) gate is an inverted AND function. Each low output of the AND function is made high, and each high output is made low. The output is low only if both inputs are high. The NAND gate is represented by the AND symbol followed by a small circle indicating an inversion of the output. See Figure 11-54.

The NAND gate, like the NOR gate, is a universal building block of digital logic. It is usually used in conjunction with other elements to implement more complex logic functions. NAND gates are also available in quad IC packaging.

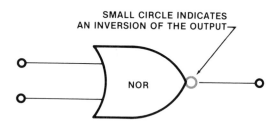

Figure 11-53. The NOR gate is represented by the OR gate symbol followed by a small circle that indicates an inversion of the output.

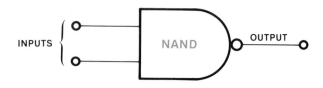

Figure 11-54. The NAND (NOT-AND) gate is an inverted AND function. Each low output of the AND function is made high and each high output is made low. The output is low if and only if both inputs are high.

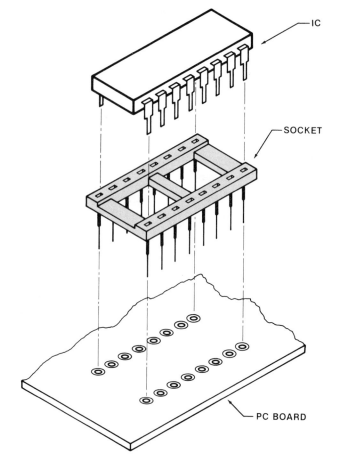

Figure 11-55. Some equipment manufacturers use sockets to hold and connect ICs on PC boards.

IC DATA SHEETS

Both the digital IC and the linear IC data sheets usually provide the following information:

1. A drawing showing the physical appearance of the package
2. A general description of the IC
3. The internal circuit schematic
4. The pin configuration of the IC
5. The normal and maximum ratings
6. The typical performance curves

The important section of IC specifications is the troubleshooting section. The section is usually entitled, "Typical Operation" or "Normal Electrical Characteristics." It is used instead of the maximum ratings because the device does not exceed the maximum ratings under normal operating conditions.

Other areas that could provide valuable trouble-shooting information are:

1. Wiring diagram
2. Pin connections
3. DC supply voltage
4. Total supply current
5. Operating temperature (ambient).

SOCKETS

Some equipment manufacturers use sockets to hold and connect ICs on PC boards. See Figure 11-55. Sockets make removing and replacing ICs easy compared to desoldering and soldering.

A screwdriver or holding bar must be used to remove ICs from sockets that are locked into position. See Figure 11-56. When the contacts are open, an IC package can be inserted or removed with little or no resistance. The unit moved to the closed position puts force on both sides of the circuit leads and locks it in place.

The two major disadvantages of sockets are that they add another component, which may fail, and they add cost to the assembly.

DIP IC REMOVAL

A DIP IC should not be removed from its socket by pulling it out by hand. This results in broken or badly bent pins. The proper procedure for removing a DIP

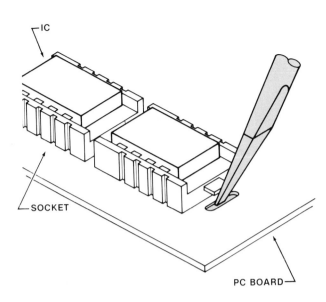

Figure 11-56. Some ICs are locked into position. These units must be unlocked with a screwdriver or a holding bar.

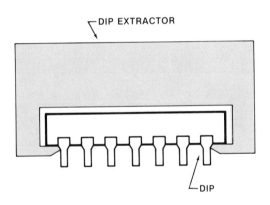

Figure 11-57. The best procedure for removing a DIP IC is to use a DIP extractor.

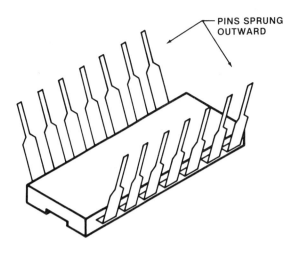

Figure 11-58. Replacement DIP ICs are shipped with their pins sprung slightly outward.

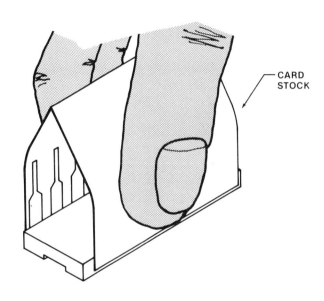

Figure 11-59. If the pins do not exactly match the socket holes, they must be bent slightly along both sides of the IC. If they are not bent evenly, a pin, or pins, may fold outward or under.

IC is to simultaneously pry loose both ends. However, the best procedure is to use one of the commercially available *DIP extractors*. See Figure 11-57. The DIP extractor securely holds the body of the IC without putting stress on any particular section. Therefore, the IC can be removed without bending any of its pins.

DIP IC REPLACMENT

Replacement DIP ICs are shipped with their pins sprung slightly outward. See Figure 11-58. This allows them to press tightly against the socket wipers when they are installed. If the pins do not exactly match the socket holes, they must be bent slightly along both sides of the IC. See Figure 11-59.

CAUTION: If they are not bent evenly, a pin, or pins, may fold outward or under.

LARGE SCALE INTEGRATION (LSI)

A large scale integration (LSI) is a chip that contains gates, amplifiers, and other circuits, and is wired as a functioning subsystem. See Figure 11-60. LSIs are used in computers and microprocessors.

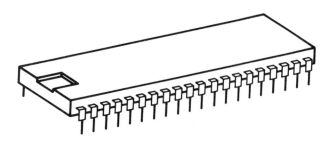

Figure 11-60. A large-scale integration (LSI) is a chip that contains gates, amplifiers, and other circuits. It is wired as a functioning subsystem.

VERY LARGE SCALE INTEGRATION (VLSI)

A *very large scale integration (VLSI)* is used to increase packing density. A VLSI contains thousands of individual devices that perform as a system.

Design and production of VLSIs are very complex. Computer-aided drafting and computer-aided manufacturing (*CAD/CAM*) techniques have become common practice in industry. VLSIs are used in electronic, automotive, and other industrial applications.

Chapter 11 - Review Questions

1. What is the main feature of an IC that makes it more popular than individual components?
2. What is the most widely used IC packaging configuration?
3. In the area of amplifiers, what are ICs mostly used for?
4. ICs usually require what type of power supplies?
5. Name the three main internal parts of an OP amp.
6. What are the two major inputs to an OP amp?
7. Which inputs are used for open-loop and closed-loop control?
8. What effect can be created when no phase compensation is provided?
9. Define nulling.
10. What is the effect of slew rate on bandwidth?
11. What can a voltage-to-current converter be used for?
12. What is a differential OP amp?
13. What is a comparator?
14. What is an active filter?
15. Name the major elements of a 555 timer.
16. How does a monostable multivibrator operate?
17. How does an astable multivibrator operate?
18. Define digital logic.
19. What are gates?
20. Where can technical data and operating characteristics of an IC be obtained?

12 FIBER OPTICS

Fiber optics is a technology that uses a thin flexible glass or plastic optical fiber to transmit light. Fiber optics is most commonly used as a transmission link. As a link, it connects two electronic circuits consisting of a transmitter and a receiver. See Figure 12-1. The central part of the transmitter is its source. The source consists of a light emitting diode (LED), infrared emitting diode (IRED), or laser diode, which changes electrical signals into light signals. The receiver usually contains a photodiode that converts light back into electrical signals. The receiver output circuit also amplifies the signal and produces the desired results, such as voice transmission or video signals.

Key Words

Absorption	Critical angle	Optical cavity
Angle of incidence	Electromagnetic interference (EMI)	Propagation
Angle of refraction	Fiber optics	Quantum efficiency
Attenuation	Isolation surge voltage	Refraction
Bar code	Light-activated device	Refractive index
Carrier wave	Normal	Scattering
Cladding	Numerical aperture	Spectral
Core		

ADVANTAGES OF FIBER OPTICS

Fiber optics systems offer many advantages for transmitting signals compared to individual wire, twisted pairs of wires, and coaxial cable. Some advantages are:

1. Large bandwidth. Since the *carrier wave* for fiber optic signals is light, fibers have bandwidths approaching 2 GHz/kM, which allows for high-speed transfer of data. This is many times higher than the highest radio frequencies.

2. Low loss (*attenuation*). Fibers can provide lower attenuation (loss) than copper wire. In addition, fibers do not react to changes in frequency, which at high frequency, can attenuate signals in metal conductors.

3. *Electromagnetic Interference (EMI)* immunity. EMI is the signal created in a conductor due to the presence of electromagnetic fields. Because fiber is an insulator, it is not affected by magnetic fields and the resulting EMI.

4. Small size. Fiber optic cable is smaller than copper cable. A small fiber optic cable has the same

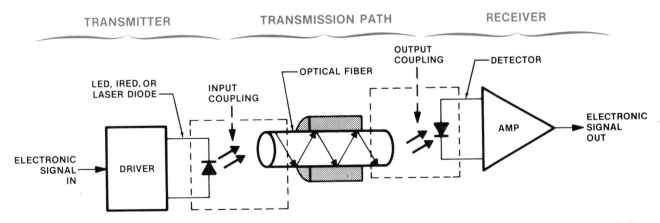

Figure 12-1. Fiber optics are used as a transmission link. The link connects two electronic circuits consisting of a transmitter and a receiver.

information-carrying capacity as a 900-pair copper cable. See Figure 12-2. Therefore, it is possible to use small conduits for optic fiber. It is also possible to increase the capacity of existing conduits. The latter is an important consideration to take when considering overcrowded conduits running under city streets.

5. Light weight. Glass and plastic weigh considerably less than copper.

6. Security. It is virtually impossible to tap a fiber optic cable without the tapping being detected. Any attempt to enter the cable will affect the quality of transmission. Also, fibers do not radiate energy, and all eavesdropping techniques are useless.

7. Safety and electrical isolation. The insulating property of fibers isolate fibers electrically. In addition, there is no spark hazard, and it can be used in flammable environments and other hazardous locations where electrical codes might prohibit the use of other methods.

NATURE OF LIGHT

When the word "light" is used to describe fiber optic operation instead of "electromagnetic radiation," it refers not only to visible light, but also to the entire spectrum where silicon-sensitive devices respond. The word "light" is incorrect because of the presence of infrared. However, it has become accepted usage in fiber optics.

Light consists of electromagnetic radiations, which are identical to the electromagnetic radiation used in radio transmission. The major difference is frequency. Light waves are composed of frequencies that are much higher than those in radio transmissions. Therefore, the wavelength of visible light frequencies are much shorter than the highest frequency generated by radio equipment. See Figure 12-3.

Two areas of the spectrum used in fiber optics are visible light and infrared light. Visible light has wavelengths from about 400 nanometers for ultraviolet to 750 nanometers for red. Nanometer (nm) means one/one millionth (1/1,000,000) of a meter. The wavelengths near the infrared region are more useful for fiber optics and extend from 750 nanometers to 1,500 nanometers. This region of the light spectrum is more useful because fibers propagate the light of these wavelengths more efficiently. *Propagation* is the speed at which light travels through a substance.

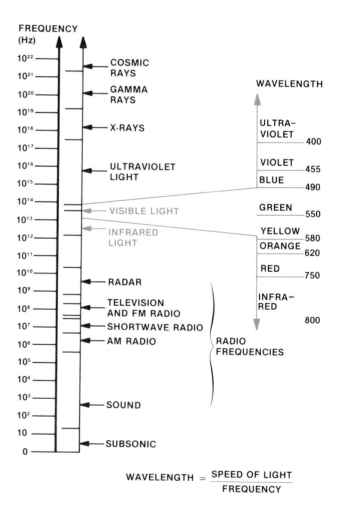

$$\text{WAVELENGTH} = \frac{\text{SPEED OF LIGHT}}{\text{FREQUENCY}}$$

Figure 12-3. Light consists of electromagnetic radiations that are composed of many different frequencies.

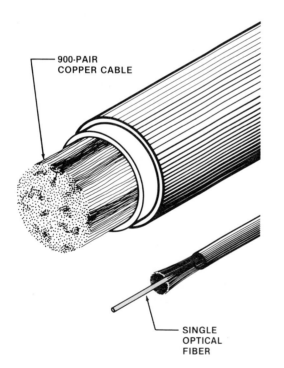

Figure 12-2. A single optical fiber has the same information-carrying capacity as a 900-pair copper cable.

Speed of Light

Light travels in free space (air) at approximately 300,000 kilometers per second (approximately 186,000 miles per second). Light travels more slowly in other substances, such as glass or water. A good example of the change in the speed of light is in a situation where light travels from air into water. See Figure 12-4. When light passes from air to water, it changes speed. This change in speed causes a deflection of light called *refraction*.

Also, different wavelengths of light travel at different speeds in the same substance. This principle can be seen in a glass prism. See Figure 12-5. White light entering a prism is composed of all colors representing different wavelengths. The prism refracts the light because each wavelength changes speed differently. Thus, the light emerging from the prism is separated, and it divides into colors of the visible spectrum.

Refraction

Knowledge of refraction of light helps explain how fiber optic cables move light through them. Three important terms associated with refraction are *normal, angle of incidence,* and *angle of refraction.* Figure 12-6 illustrates the interrelationship of these terms. The normal is an imaginary line perpendicular to the interface of two materials, such as glass and air. The angle of incidence is the angle formed by the incoming ray (incident ray) and the imaginary line (normal). The angle of refraction is the angle formed between the normal and the refracted ray.

The angle of incidence determines what happens to the incident ray. As the angle of incidence increases,

the chances of this incident ray entering the second material decreases. The angle at which this takes place is the *critical angle.* The critical angle varies with different substances. Figure 12-7 shows what happens when the critical angle is reached. The incident ray reaches a 90° refraction and does not enter the second

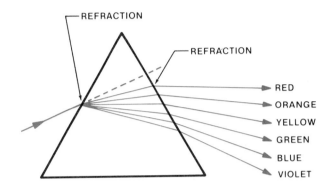

Figure 12-5. Light of different wavelengths traveling at various speeds through a prism is refracted differently.

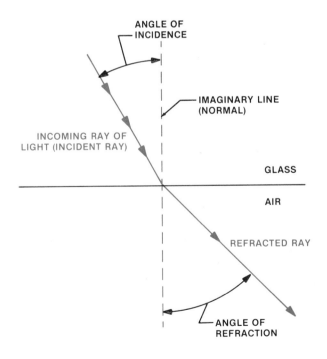

Figure 12-6. The angle of incidence is the angle formed by the incoming ray (incident ray) and the imaginary line (normal). The angle of refraction is the angle formed between the normal and the refracted ray.

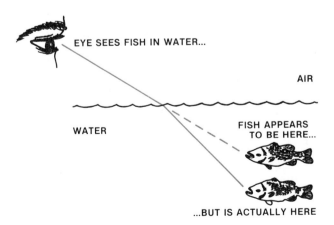

Figure 12-4. When light passes from air into water, it changes speed. This change causes a deflection of light called refraction.

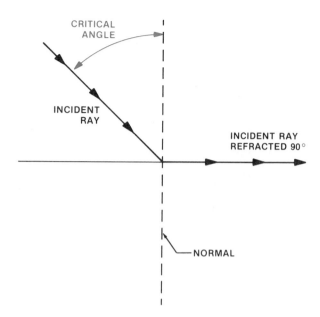

Figure 12-7. The angle at which light no longer passes into the second material is the critical angle.

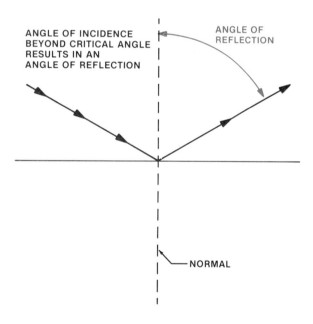

Figure 12-8. When the angle of incidence passes beyond the critical angle, the light is totally reflected.

OPTICAL FIBER

Figure 12-9 shows the typical construction of an optical fiber. The *core* is the actual path for light. Although

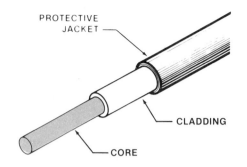

Figure 12-9. Optical fibers consist of a core, cladding, and protective jacket.

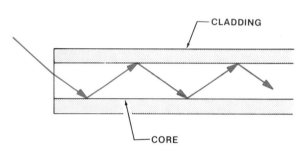

Figure 12-10. Light is reflected through the fiber core by total internal reflection.

the core occasionally is constructed of plastic, it is typically made of glass. A *cladding* layer, usually of glass or plastic, is bonded to the core. The cladding is enclosed in a jacket for protection.

Reflection Within an Optical Fiber

A basic optical fiber forms two concentric layers with the core and cladding. See Figure 12-10. The core (inner layer) has a higher refractive index than the cladding (outer layer) has. The *refractive index* indicates the speed of light through a substance compared to that of free space. Since the core has a higher refractive index, light moves more slowly through the core than through the cladding.

Any light injected into the core that strikes the surface at an angle greater than the critical angle is reflected back into the core. Since the angles of incidence and reflection are equal, the ray of light continues zigzagging down the length of the core by total internal reflection. This light is trapped in the core. Light striking the surface at less than critical angle passes into the cladding and is lost. The total amount of light moving through the fiber is determined by fiber size, fiber construction, and the nature of the light injected.

material. If the angle of incidence increases past the critical angle, the light is totally reflected back into the first material. See Figure 12-8.

Cables

For most applications, optical fiber must be cabled by enclosing a fiber, or fibers, within a protective structure. Cabling can range from application of a simple insulating jacket to the assembly of jackets, tubes, and strengthening members within a multifiber cable. See Figure 12-11. Cable components must protect the fiber throughout its lifetime. They are especially important during installation when greater stress is likely to be applied.

LIGHT SOURCES

The light source for fiber optics is usually a light emitting diode (LED) or a laser diode. The LED, compared to a laser diode, provides less power and operates at slower speeds. However, the LED is suitable for most applications requiring transmission distances up to several kilometers, and speeds up to several hundred megabits per second. The LED is also more reliable, less expensive, and easier to use than the laser diode.

Light Emitting Diode (LED)

The light emitting diode (LED) is a PN junction diode that emits light when forward biased. See Figure 12-12. The light emitted can be either invisible (infrared), or it can be light in the visible spectrum. LEDs for electronic applications, due to the spectral response of silicon and efficiency considerations, are usually infrared emitting diodes (IRED).

The IRED is an LED that emits invisible light near the infrared region of the light spectrum. Generally, gallium arsenide and gallium arsenide phosphide are the materials used for the IRED. They are housed in either metal or plastic. See Figure 12-13. The electrical characteristics of the IRED are similar to those of any other PN junction diode. The IRED has a slightly

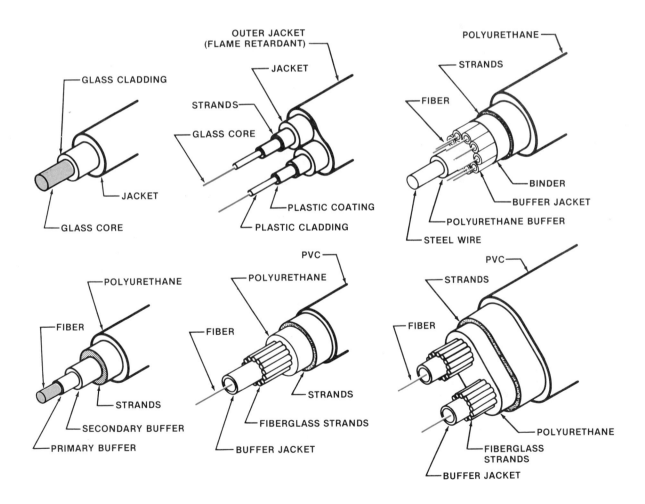

Figure 12-11. Cabling can range from application of a simple insulating jacket to the assembly of jackets, tubes, and strengthening members within a multifiber cable.

higher forward voltage drop than a silicon diode due to the higher energy necessary. It has a fairly low reverse breakdown voltage due to the doping levels required for efficient light production. The light output of the diode is dependent upon power supplies and is measured in milliwatts.

Testing LEDs. The output of an LED or IRED is usually determined by using a test fixture. See Figure 12-14. Electrically, the LED and IRED are powered and monitored by a circuit similar to the one shown in Figure 12-15. The object of the test is to determine the efficiency of the LED or IRED. If the fixture has been properly calibrated, it will measure the efficiency of the LED or IRED based on the amount of input current needed to drive the emitter, versus the amount of output current derived from the photovoltaic cell (solar cell).

In an ideal emitter-receiver situation, such as the test fixture, every photon generated by the emitter should result in an electron being generated in the receiver. This ideal situation would result in a *quantum efficiency* ratio of 1. Quantum efficiency is a method of calculating the number of electrons produced for each photon striking the semiconductor. Mathematically, it is expressed as:

$$quantum \ efficiency = \frac{electrons}{photons}$$

For all semiconductor devices, the quantum efficiency is less than 1.

Mounting an LED. The LED may be held in position with a lock collar/bushing arrangement. See Figure 12-16. It may also be held in position by press-fitting the LED into place. See Figure 12-17.

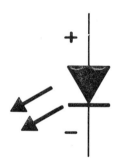

Figure 12-12. The LED is a PN junction diode that emits light when forward biased.

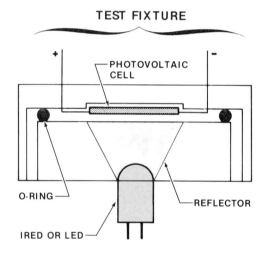

Figure 12-14. The output of an LED or IRED is determined by using a test fixture.

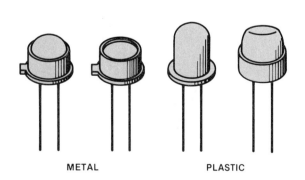

Figure 12-13. LEDs are housed in either metal or plastic.

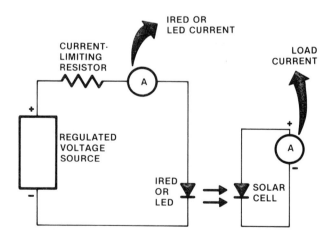

Figure 12-15. When properly calibrated, the amount of input current (LED or IRED current) and output current (load current) can be measured to determine the efficiency of the LED or IRED.

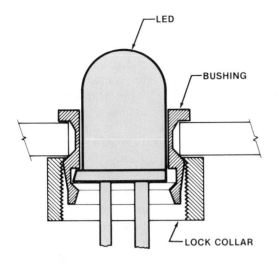

Figure 12-16. LEDs may be held in position with a lock collar/bushing arrangement.

CAUTION: When press-fitting, do not use force on the LED leads. Pressure should be applied carefully to the body of the LED with needle-nose pliers or a nut driver.

Laser Diode

The laser diode is different from an LED in that it has an *optical cavity*, which is required for lasing production (emitting coherent light). The optical cavity is formed by coating opposite sides of a chip to create two highly reflective surfaces. See Figure 12-18.

At low-drive currents, the laser diode operates like an edge-emitting diode, spontaneously emitting light from its side (edge). At a threshold current level, laser action begins, and the chip emits coherent, or nearly coherent, light. The laser relies on the injection of high current densities into the optical cavity. Photons created by recombination are partially trapped within the optical cavity because some reflect against the mirrorlike coated sides.

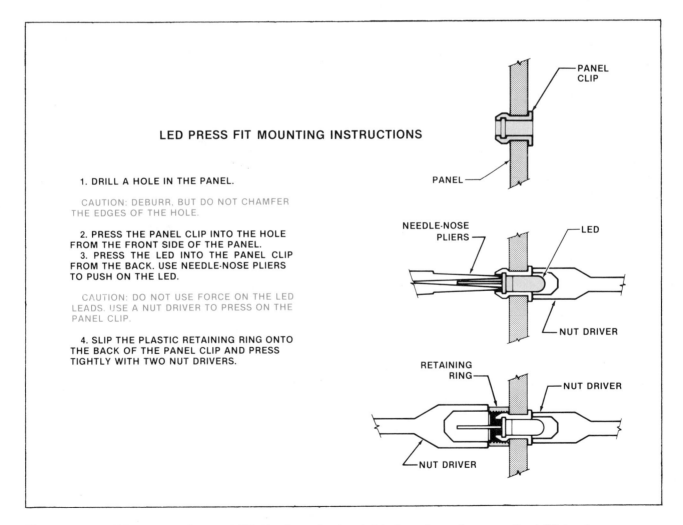

Figure 12-17. When mounting an LED, the important point is to not use force on the LED leads.

If a photon strikes an excited electron, the electron immediately recombines, releasing a photon of the same frequency, direction, and phase as the incident photon. In other words, the incident photon has stimulated the emission of an identical photon. While a portion of the light remains trapped, by stimulating further emission, the rest of the light escapes through the two coated sides as intense beams. Figure 12-19 shows typical curves of output power versus drive current for an LED and a laser.

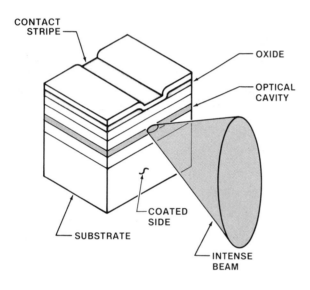

Figure 12-18. Laser action begins at threshold current level, and the chip emits coherent, or nearly coherent, light. This light is in the form of an intense beam.

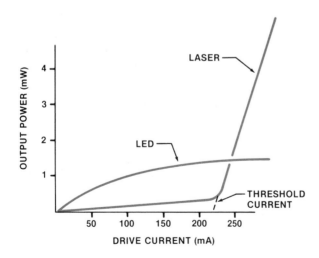

Figure 12-19. The graph shows the output power versus drive current for an LED and a laser.

ATTENUATION

Attenuation is loss of power. During transit, light pulses lose some of their photons, thus reducing their amplitude. Attenuation for an optical fiber is usually specified in decibels per kilometer (dB/km). For commercially available optical fibers, attenuation ranges from 1 decibel per kilometer for premium small-core glass fibers to over 2,000 decibels per kilometer for large-core plastic fibers.

Absorption

Glass fibers are made of ultrapure glass that has purity exceeding that of semiconductors. If sea water were as pure as a fiber glass, the bottom of the deepest part of the Pacific Ocean could be seen from the surface of the ocean. Some impurities still remain as residues in glass and other dopants that are added purposely to obtain certain optical qualities. Unfortunately, impurities in glass absorb light energy, turning photons into heat. See Figure 12-20.

Scattering

Scattering results from imperfection in fibers and from the basic structure of fiber. Unintentional variation in the density and in the geometry of the fiber occurs during fiber manufacturing and cabling. Small variations in the core diameter, microbends, and irregularities (bubbles) in the core-to-cladding interface cause loss. See Figure 12-20. The angle of incidence of rays striking such variations significantly change some of the rays. They are refracted onto new paths and are not subject to total internal reflection.

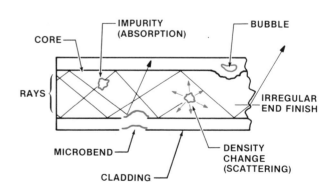

Figure 12-20. Attenuation, or loss, can be caused by impurities or imperfections in the fiber optic cable.

Wavelength

Fiber attenuation is closely related to wavelength, and requires a careful balance between light sources and fibers. Manufacturers specify attenuation of a fiber at a certain wavelength. Most manufacturers also provide a curve showing attenuation as a function of wavelength. This data should be used when matching a fiber to a light source.

Numerical Aperture

The *numerical aperture* is a measure of the light of a fiber gathering ability. Unless light is placed into the fiber at the proper angle, rays will be lost. Light rays

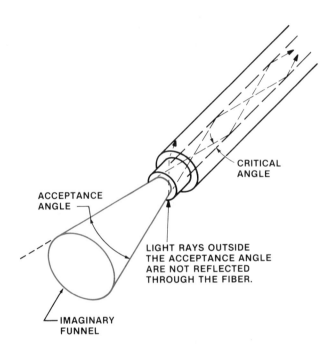

Figure 12-21. Rays outside the acceptance angle (not within the imaginary funnel) are not reflected and are lost.

must enter and strike the cladding at an angle greater than the critical angle. Figure 12-21 shows an imaginary funnel formed at the entrance of the fiber. Any rays striking within this area will travel through the fiber. Rays outside the area will strike the cladding at less than the critical angle and will be lost. Loss created by misalignment is always a factor to consider.

FIBER COUPLING

The ideal interconnection of one fiber to another is an interconnection that has two fibers that are optically and physically identical. These two fibers are held together by a connector or splice that squarely aligns them on their center axes. The joining of the fibers is so nearly perfect that the interface between them has no influence on light propagation. Such a perfect connection will always be limited by two factors:
  1. Variations in fibers.
  2. Tolerances required in the connector or splice make their manufacture impractical.
These two factors affect cost and ease of use.

Fiber Coupling Hardware

For an electrician, there is little that can be done about the design of coupling materials available. However, proper installation procedures should be understood and followed. Splices and fiber interconnections are often more of a negative factor than poor quality materials are because of alignment problems that can arise. Figure 12-22 shows three common errors encountered in coupling. The elimination of these problems can be accomplished through proper installation of fiber splices, connectors, and couplers.

Splices. The square-tube splice is an alignment mechanism that uses an internal V-groove to achieve two points of contact. See Figure 12-23. The fibers are installed in a square tube filled with epoxy. Maintaining a slight bend on the fibers, as they are pushed into the tube, forces their ends into a V-groove formed by

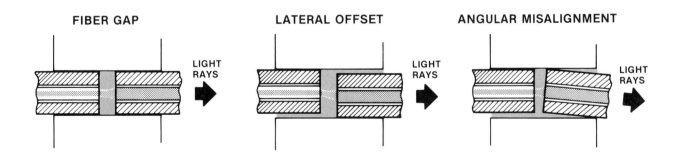

Figure 12-22. Three major areas of attenuation or loss are fiber gap, lateral offset, and angular misalignment.

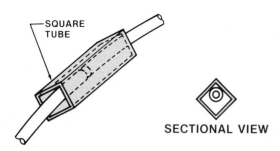

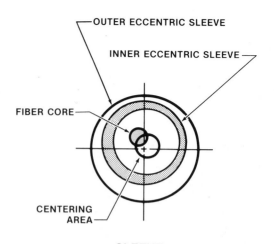

Figure 12-23. The square-tube splice is an alignment mechanism using an internal V-groove to achieve two points of contact.

Figure 12-25. Double-eccentric connectors use eccentric sleeves (inner and outer) to align fibers.

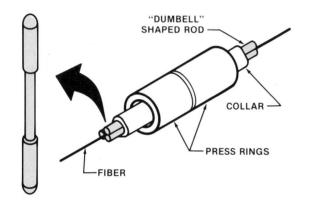

Figure 12-24. With the three-rod splice, press rings are forced onto the raised portion of the collar, pressing inward on the dumbell-shaped rods to hold the fibers.

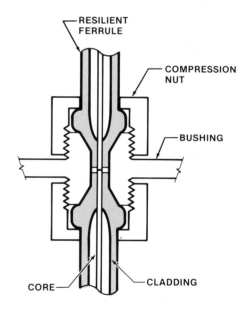

Figure 12-26. With a resilient-ferrule connector, the compression forces both fibers to align to a common axis.

the corners of the tube. The splice works well with fibers of nearly the same diameter.

Another type of splice is the three-rod splice. See Figure 12-24. The splice uses three dumbell-shaped rods enclosed in a collar. The collar is raised slightly in the center. The fibers fit easily between the rods. When press rings are forced onto the raised portion of the collar, the rods press inward to hold the fibers. The resiliency of the collar and the inward bend of the rods permit compensation for differences in fiber diameters.

Connectors. Eccentricity is one of the problems encountered when working with fibers. Eccentricity exists when the fiber is not a perfect circle and does not properly align with another fiber. Double-eccentric connectors are used to correct this problem. See Figure 12-25. Fibers are held by two eccentric sleeves (outer and inner), one in each connector half. The sleeves are rotated to bring the fiber axes into very close alignment. Then, the eccentric sleeves are locked into place. Such tuning of a connector results in good

performance. However, it may require test equipment to indicate peak performance.

Another popular fiber-optic connector is the resilient ferrule. It uses a secondary alignment mechanism. See Figure 12-26. Fibers are held within resilient ferrules that are mated in a bushing which tapers inward from both ends. As the ferrules are pressed into the bushing, the tapers compress them. The compression forces both fibers to align to a common axis. Because of the compression, the hole in the ferrule can be made to fit the largest fiber diameter within tolerance. During termination, when the fiber is epoxied in the ferrule, the ferrule is pressed tightly against the fiber by a compression nut. By changing the hole sizes in the ferrule, the connector can handle any size of fiber 125 μm and larger.

The resilient ferrule has two main advantages. First, it aligns fibers on a common axis, regardless of their diameter variations. Second, the concept is simple and straightforward enough to reduce tolerance buildup effects. The result is a connector that is inexpensive compared to other styles of connectors with comparable performance. The disadvantage is the bonding of the fiber to the connector. Epoxy curing and polishing are time consuming.

Couplers. Fiber optic splices and connectors are used for point-to-point applications that transport light from source to detector. Couplers allow light to be distributed among several fibers. Coupling is desirable in bus-structure applications or in distributed networks. The two main types of couplers are the T-coupler and star coupler.

T-coupler. The T-coupler is an in-line device used for tapping a main optical bus. One approach to tapping the light is to use lenses and a beamsplitter. See Figure 12-27. A beamsplitter is a coated plate that reflects part of the light and transmits the rest. The structure of the beamsplitter determines the reflection-to-transmission ratio. Loss of optical signal power increases linearly with the number of couplers in series with one another. Not only is there an insertion loss associated with each coupler, but the light is also divided for each. Figure 12-27 shows the internal operation of a T-coupler, as well as its application in a network.

Star Coupler. Star couplers offer parallel arrangements. See Figure 12-28. As transmitted light enters

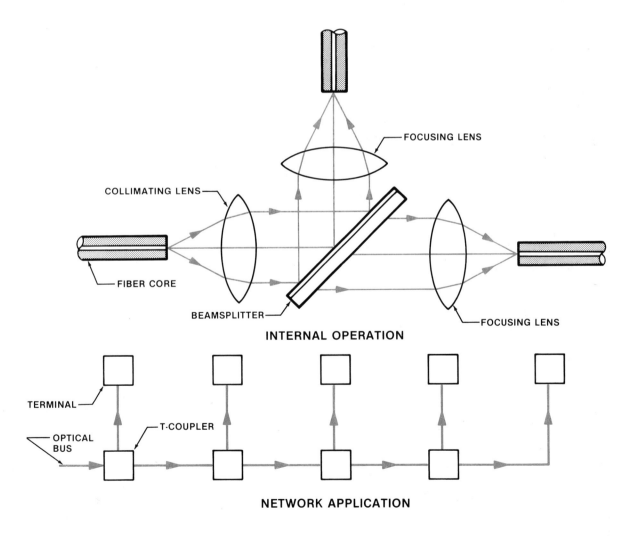

Figure 12-27. A T-coupler allows in-line tapping of a main fiber (optical) bus using lenses and a beamsplitter.

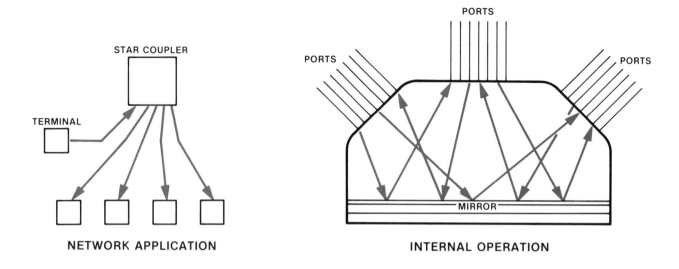

Figure 12-28. The advantage of a star coupler over a T-coupler is that there is only one coupling loss associated with dividing light from one fiber into several fibers.

a star coupler through its input ports, the light spreads out and strikes a mirror that reflects it into the output ports. The advantage of a star coupler over a T-coupler is that there is only one coupling loss associated with dividing light from one fiber into several fibers. In addition, as the number of output ports increases, the loss from beamsplitting increases only slightly in relation to the number of ports.

Splitter/Combiner. T-couplers and star couplers have relatively high losses, usually greater than 5 decibels. Another approach to distributing a signal is with a splitter/combiner that uses resilient-ferrule connectors. Consider a source that provides a uniformly distributed large diameter spot of light at its output. Connecting a small fiber to it allows a large portion of the available light to go unused simply because the fiber core is smaller than the active window of the source. To eliminate this loss, seven fibers can be packaged into a single resilient-ferrule connector to provide a large effective active area covered by the cores of the fibers. See Figure 12-29. Each fiber now receives nearly the same amount of light as a single fiber connected to the larger source would. The other end of each fiber is terminated in a separate connector. The light is distributed into the seven fibers. The opposite is also possible. Light can be combined to a common detector.

Losses within a splitter/combiner are comparable to those found in connectors. It is less costly than using a separate T-coupler or star coupler since it eliminates the expense of the coupler itself.

LIGHT-ACTIVATED DEVICES

Once light rays have passed through the optical fiber, they must be detected and converted back into electrical signals. The detection and conversion is accomplished with *light-activated devices*, such as PIN photodiodes, phototransistors, photo SCRs, and phototriacs.

PIN Photodiode

Figure 12-30 shows a typical construction and operation of the PIN photodiode. (PIN stands for P-type material, insulator, and N-type material, respectively.) The operation of a PIN photodiode is based on the principle that light radiation, when exposed to a PN junction, momentarily disturbs the structure of the PN junction. The disturbance is due to a hole created when a high-energy photon strikes the PN junction and causes an electron to be ejected from the junction. Thus, light creates electron-hole pairs, which act as current carriers.

Electrical Considerations of PIN Photodiode. When connected in a circuit, the PIN photodiode is placed in series with the bias voltage and connected in reverse bias. See Figure 12-31. A PIN photodiode is mounted to the base of a transistor-type header, which is then sealed with a plane glass or plastic window. It is then capped to form a hermetically sealed package. See Figure 12-32.

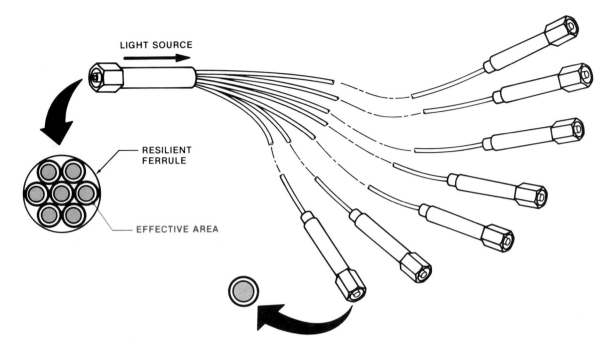

Figure 12-29. To eliminate loss of light, seven fibers can be packed into a single resilient-ferrule connector to provide a large effective area.

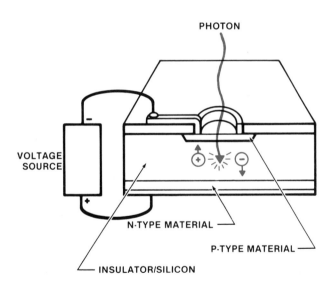

Figure 12-30. The operation of a PIN photodiode is based on the principle that light radiation, when exposed to a PN junction, momentarily disturbs the structure of the PN junction. This creates electron-hole pairs.

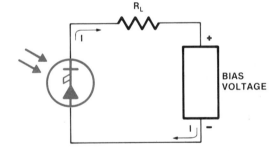

Figure 12-31. When placed in a circuit, the PIN photodiode must be properly biased and connected in series with the biasing voltage.

PIN Photodiode Alignment. For best performance of a PIN photodiode, the light must be focused on the PN junction. The reason for this is that the photocurrent produced is the result of the collection, at the PN junction, of the minority carriers freed near the junction by the photoelectric process. The electrical field in the crystal is negligible except near the PIN junction. Most of the carriers reach the junction by diffusion. The farther away from the junction that the minority carriers are freed, the less chance they have of being collected. Thus, the response of the PN photodiode decreases on either side of the junction, as shown in the curve of Figure 12-33. In some applications, a lens is provided to focus the light on the sensitive area of the PIN photodiode.

Spectral Response of PIN Photodiode. The absorption of light in silicon decreases with increasing radiation wavelength. Therefore, as the radiation wavelength decreases, a larger percentage of the electron-hole pairs are created closer to the silicon surface. This results in the PIN photodiode exhibiting a peak-response point at some radiation wavelength. At this wavelength, a maximum number of electron-hole pairs are created near the collector-base junction. A *spectral response curve* is used to determine the optimum wavelengths for a PIN photodiode. See Figure 12-34. A spectral response curve expresses the quantum efficiency of the PIN photodiode in relation to the wavelength of light placed upon it. An ideal quantum efficiency would be 1, in which every photon creates one electron. From the information on the graph, it is seen that none of the light sources produces a quantum efficiency of 1. However, gallium arsenide approaches .6. Also, the spectral response for this device ranges from 400 to 1,100 nanometers.

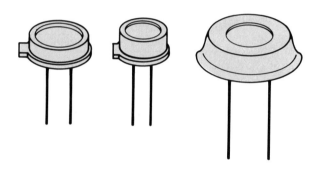

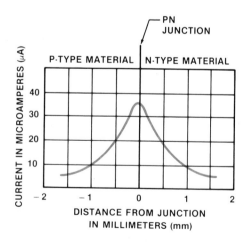

Figure 12-32. A PIN photodiode is mounted to the base of a transistor-type header, which is then sealed with a plane glass or plastic window to form a hermetically sealed package.

Figure 12-33. For the best response of a PIN photodiode, the light must be focused directly on the PN junction.

Advantages and Disadvantages of PIN Photodiodes.

PIN photodiodes can respond quickly to changes in light intensity. They are very useful for applications in which light changes at a rapid rate. The major disadvantage is that the output photocurrent is relatively low in relation to other photoconductive devices.

PIN photodiodes are small and efficient. They have low operating voltages, low power consumption, low noise, and simple circuitry. Therefore, they have the broadest range of applications.

Phototransistor

A phototransistor combines the effect of the photodiode and the switching capability of a transistor. Schematically, the phototransistor may be represented by either one of the symbols shown in Figure 12-35. Electrically, the phototransistor, when connected in a circuit, is placed in series with the bias voltage so that it is forward biased. See Figure 12-36.

With a two-lead phototransistor, the base lead is replaced by a clear covering. This covering allows light to fall on the base region. See Figure 12-37. Light falling on the base region causes current to flow between emitter and collector. The collector-base

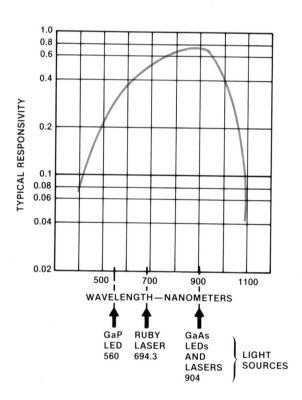

Figure 12-34. A spectral response curve is used to determine the optimum wavelength for a PIN photodiode.

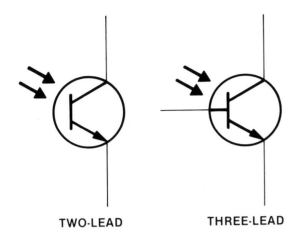

TWO-LEAD   THREE-LEAD

Figure 12-35. A phototransistor combines the effect of the photodiode and the switching capability of a transistor.

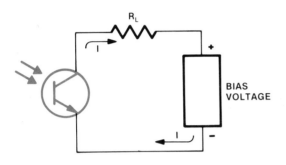

Figure 12-36. The phototransistor must be properly biased and placed in series with the biasing voltage.

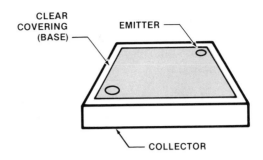

Figure 12-37. In a two-lead phototransistor, the base lead is replaced by a clear covering that allows light to fall on the base region of the transistor.

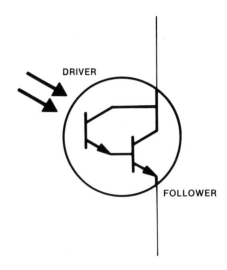

Figure 12-38. The photodarlington is a multistage amplifier in one package. It works on the same principle that a single phototransistor does.

junction is enlarged and works as a reverse-biased photodiode controlling the phototransistor. The phototransistor conducts more or less current depending upon the light intensity. If light intensity increases, resistance decreases, and more emitter-to-base current is created. Although the base current is relatively small, the amplifying capability of the small base current is used to control the larger emitter-to-collector current. The collector current depends on the light intensity and the DC current gain of the phototransistor. In darkness, the phototransistor is switched OFF with the remaining leakage current. This remaining leakage current is called collector dark current.

Advantages and Disadvantages of Phototransistors. Phototransistors have an advantage over photodiodes in that they have a sensitivity approximately 50 to 500 times greater. Two distinct disadvantages are that they are sensitive to temperature changes, and that protection against moisture is required.

Although the phototransistor can produce a higher output current than a photodiode, the phototransitor loses some of its response speed. Thus, photodiodes are still often used when phototransistors cannot meet the response time. Phototransistors do very well in applications such as smoke detectors, counters, photographic meters, and mechanical positioning systems where speed is not critical.

Photodarlington. The photodarlington works on the same principle as a phototransistor does. The collector-base junction of the driver transistor is radiation-sensitive, and it controls the driver transistor. The driver transistor controls the follower transistor. See Figure 12-38. The darlington configuration yields a high current gain, which results in a photodetector with very high light sensitivity.

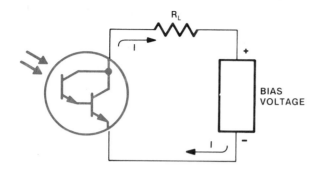

Figure 12-39. A properly biased photodarlington can be used where light to be detected is of a low level.

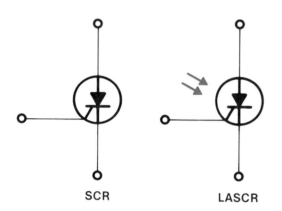

SCR   LASCR

Figure 12-40. The schematic of an LASCR is identical to the schematic of a regular SCR. However, arrows are added in the LASCR schematic to indicate a light-sensitive diode.

Photodarlington circuits are usually formed simultaneously and packaged as a single device. Although the photodarlington is more sensitive than the photodiode or phototransistor, it has a slower response to changes in light intensity. A properly biased photodarlington is shown in Figure 12-39. The photodarlington is useful where the light to be detected is of a low level.

Light Activated SCR (LASCR)

The schematic of a light activated SCR (LASCR) is identical to the schematic of a regular SCR. The only difference is that arrows are added in the LASCR schematic to indicate a light-sensitive device. See Figure 12-40.

Like the photodiode, current is of a very low level in an LASCR. Even the largest LASCRs are limited

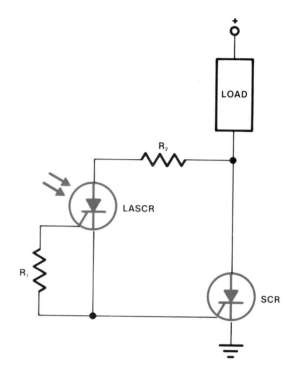

Figure 12-41. When larger current requirements are necessary, the LASCR can be used as a trigger circuit for a conventional SCR.

to a maximum of a few amps. When larger current requirements are necessary, the LASCR can be used as a trigger circuit for a conventional SCR. See Figure 12-41. In this circuit, the SCR is normally OFF since its gate circuit can be considered open with the LASCR in darkness. When the LASCR is triggered by a pulse of light, it turns ON. This in turn triggers the SCR, which supplies the heavier load current.

The primary advantage of the LASCR over the SCR is its ability to provide isolation. Since the LASCR is triggered by light, the LASCR provides complete isolation between the input signal and the output load current.

Phototriac

The gate of the phototriac is light-sensitive. It triggers the triac at a specified light intensity. See Figure 12-42. In darkness, the triac is not triggered. The remaining leakage current is called peak blocking current. The phototriac is bilateral and is designed to switch AC signals.

Optoelectronic Packaging

Optoelectronics encompasses diodes, transistors, SCRs, and triacs. Optoelectronic components require

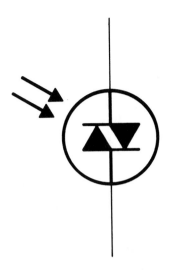

Figure 12-42. The phototriac, like the SCR, can be triggered by light radiation of a certain density.

packaging that not only protects the chip, but also allows light to pass through the package to the chip. In other words, it is a semiconductor package with a window. See Figure 12-43. Generally, the window does not include a lens as part of the package. Quality control and placement of the lens can be costly and unpredictable. Therefore, external lens and directional optic systems are usually specified for installations that require quality optics.

Another factor in packaging is the choice of a plastic or hermetic package. The hermetic package operates at higher power and over a wider temperature range. It is also more tolerant of severe environments. However, it is more expensive than plastic.

OPTOCOUPLER (OPTOISOLATOR)

There are many situations in which information must be transmitted between switching circuits that are electrically isolated from each other. This isolation has

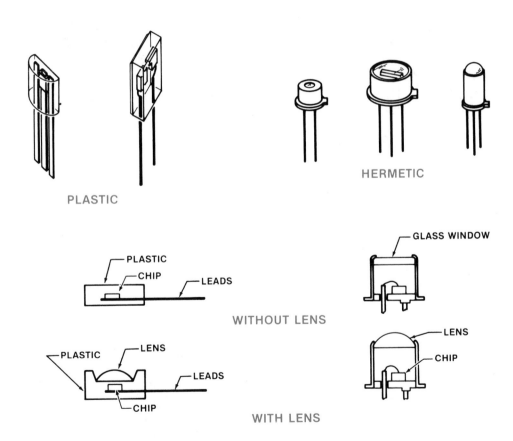

Figure 12-43. Optoelectronic components require packaging that not only protects the chip, but also allows light to pass through the package to the chip.

been traditionally provided by relays, isolation transformers, and line drives and receivers. Opto-couplers (optoisolators) can be used effectively to solve the same problems that traditional devices have. The optocoupler is mostly used in applications where high voltage and noise isolation, as well as small size, are considerations. By coupling two systems together with the transmission of radiant energy (photons), the need for a common ground is eliminated. Optocouplers are packaged as dual in-line packages (DIPS), interrupter modules, and reflector modules.

Optocoupler Construction

Externally, an optocoupler is usually constructed as a dual in-line plastic package. See Figure 12-44. An optocoupler consists of an infrared emitting diode (IRED) as the input stage, and a silicon phototransistor as the output stage. See Figure 12-45. Internally, the

optocoupler uses a glass dielectric sandwich to separate input from output. See Figure 12-46. The coupling medium between the IRED and sensor is the infrared transmitting glass. This provides one-way transfer of electrical signals from the IRED to the photodetector (phototransistor) without electrical connection between the circuitry containing the devices.

Photons emitted from the IRED (emitter) have wavelengths of about 900 nanometers. The detector (transistor) responds effectively to photons with this same wavelength. Thus, input and output devices are always spectrally matched for maximum transfer characteristics. The signal cannot go back in the opposite direction because the emitters and detectors cannot reverse their operating functions.

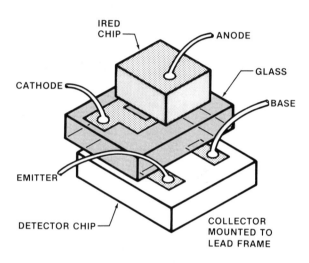

Figure 12-46. The optocoupler uses a glass dielectric sandwich to separate input from output.

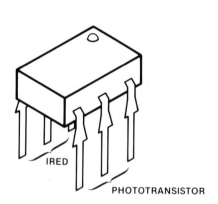

Figure 12-44. An optocoupler or optoisolator is usually constructed as a dual in-line package (DIP).

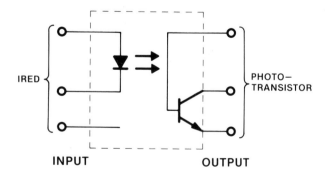

Figure 12-45. Internally, an optocoupler consists of an infrared emitting diode (IRED) as the input stage and a silicon NPN phototransistor as an output stage.

Optocoupler Isolation Voltage

The primary function of the optocoupler is to provide electrical separation between input and output, especially in the presence of high voltages. The ability of an optocoupler to provide this separation is usually expressed as an *isolation surge voltage*. It is a measure of the integrity of the package and the dielectric strength of the insulating material. The amount of electrical isolation between the two devices is controlled by the material in the light path, and the physical distance between the emitter and detector.

Although the DIP package is the most common optocoupler, other packages are available to provide higher isolation surge voltage and other special requirements. For very high isolation surge voltage requirements (10 to 50 kV), the interrupter module can be modified at a very low cost. This modification is done by putting a suitable dielectric (glass, acrylic, or silicone) in the air gap, and insulating and encapsulating the lead wires.

NOTE: Isolation surge voltage may be expressed as a steady-state isolation voltage. Steady-state isolation voltage ratings are those ratings for a working voltage that are continuously applied to the device. Isolation surge voltage is intended for short periods only and is usually higher than steady-state. Isolation surge voltage may be 1,200 volts peak where steady-state may be 900 volts peak.

Interrupter Optocoupler Module

The interrupter optocoupler module is used to eliminate the mechanical positioning problem encountered when adjusting the emitter and detector for proper sensing. See Figure 12-47. These units are constructed so that the input and output are set as a coupled pair. All alignment and distance problems have already been taken into account. The technician only needs to apply the proper power source to the device. The schematic of the interrupter optocoupler module is the same as that of the optocoupler. However, the indication of the light beam path in the interrupter module is exposed. See Figure 12-48. These types of sensors are in the same category that a

precision mechanical-limit switch is in. The activating mechanism is blocking or reflecting light instead of applying a force. See Figure 12-49.

Reflector Modules

Reflector modules are used when only one surface of an object is readily accessible. An example of such an object is an encoder wheel. See Figure 12-50. The encoder wheel is a disk with alternating reflective and dark segments arranged in tracks. By bouncing a light source against this surface, light returns when a reflective segment is present.

With a reflector module, the emitter and detector are mounted side by side and are aimed so that the light sources converge at a point just beyond the surface of the module. See Figure 12-51. As the reflective segment of the encoder wheel passes by the emit-

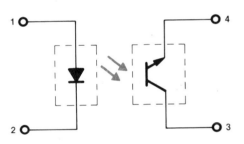

Figure 12-48. The schematic of the interrupter module is the same as that of the optocoupler. However, the indication of the light beam path in the interrupter module is exposed.

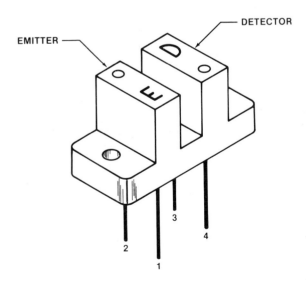

Figure 12-47. The interrupter optocoupler module is used to eliminate the mechanical positioning problems encountered in adjusting the emitter and detector for proper sensing.

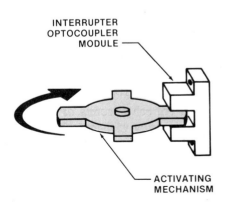

Figure 12-49. The activating mechanism for an interrupter optocoupler module is based on blocking or reflecting light instead of applying force.

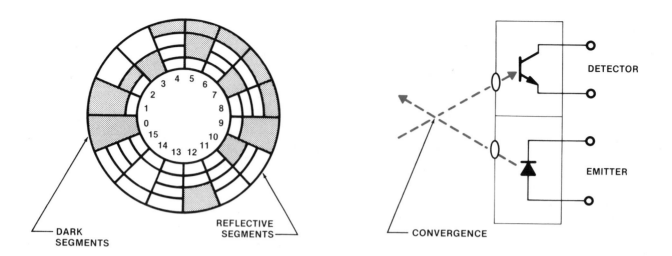

Figure 12-50. The encoder wheel is a disk with alternating reflective and dark segments.

Figure 12-51. With a reflector module, the emitter and detector are mounted side by side and are aimed so that the light sources converge at a point just beyond the surface of the module.

ter, the detector receives a return light source. When the dark segment of the encoder wheel passes by the emitter, the detector will not receive a return light source.

Reflector modules are used in robotics to help control robot arms. See Figure 12-52. Other

applications are in printers, plotters, tape drives, positioning tables, and automatic handlers. A reflector module consists of a small plastic or metal case that can be mounted in a variety of positions. See Figure 12-53.

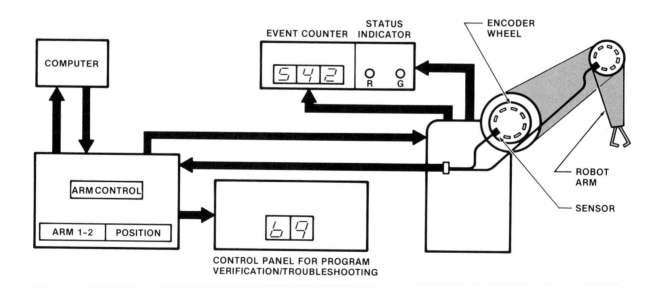

Figure 12-52. Reflector modules with encoder wheels are used in robotics to control robot arms.

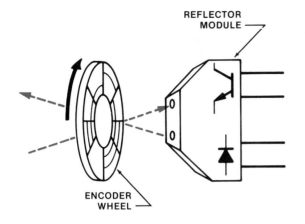

Figure 12-53. The reflector module usually consists of a small plastic or metal case that can be mounted in various positions.

BAR CODES AND BAR CODE SCANNING

Many cash registers now identify a product sold by scanning a *bar code* (line code) printed on the product label or packaging. See Figure 12-54. Bar codes are usually horizontal with alternating vertical dark lines and light spaces. Data is encoded by varying the width of these bars and spaces. To retrieve data, a wand scanner is moved across the bar code.

The wand scanner is a self-contained unit that has both a light source and a light detector. See Figure 12-54. The light source is projected through an opening in the tip of the wand. The beam of light strikes the bar code and is reflected back into the wand tip and to the light detector. The light source is connected to the tip through an optical fiber, which guides the light to the tip. The reflected light uses the same optical path from the tip back to a Y-junction. A portion of the reflected light is then directed to the light detector.

TACHOMETER PROBE

Figure 12-55 shows the physical and schematic layout of a tachometer probe using optoelectronic components. The reflective optical coupler contained in the tachometer probe consists of an IRED and a photodarlington transistor. Resistor R_1 is used to limit diode current. Resistor R_2 is the phototransistor load resistor.

If the light emitted by the LED is reflected back to the phototransistor, the phototransistor conducts and the collector voltage decreases. If the light is not reflected back, the phototransistor does not conduct, and its collector voltage remains high.

When properly installed, the output pulse of the probe frequency is proportional to the engine RPM. The output can be viewed on an oscilloscope, measured on a frequency counter, or converted to an analog RPM reading with associated circuitry.

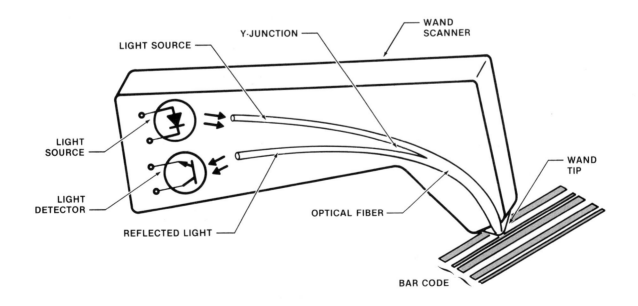

Figure 12-54. In a bar code reader, the light emitted from the source strikes the bar code and is reflected back into the wand tip and to the light detector.

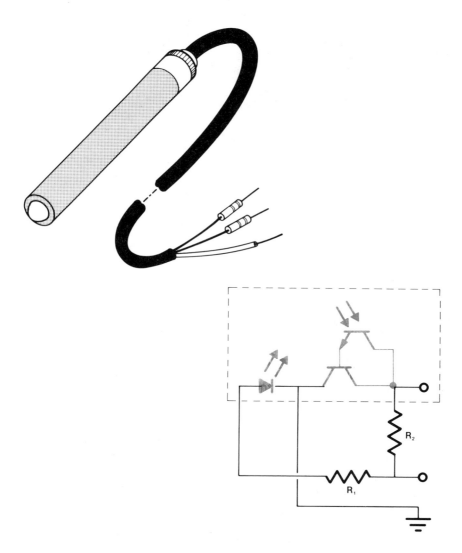

Figure 12-55. The reflective optical coupler in a tachometer probe consists of an IRED and a photodarlington transistor.

COLORIMETER

The chemical composition of certain gases and liquids can be determined by measuring the amount of a certain color light that they absorb. (A number of gas analyzers and similar apparatus are also based on this principle.) An example of this type of equipment is the colorimeter. See Figure 12-56. The term colorimeter refers to a color-sensitive meter designed for the chemical analysis of fluids.

The fluid to be analyzed passes through the flow cell as shown in Figure 12-56. A light beam is passed through the fluid and through a color filter to a photodetector. The intensity of the light beam (of the color chosen) that strikes the photodetector provides the basis for measuring the amount of light absorbed by the fluid. This in turn indicates the density. The output of the photodetector is then sent through an amplifier circuit to a recorder. The compartment that houses the light source is sealed from the fluid. This is done by a window that passes the light beam and keeps out the fluid. The compartment that houses the photodetector is similarly constructed.

HIGH-VOLTAGE SWITCHING

Light-activated phototransistors can be used as a triggering mechanism for SCRs. If the SCRs are in series, very high voltages can be safely switched. See Figure 12-57.

When the flash tube is activated, light enters the input of the fiber optic bundle and splits into ten equal outputs. Each burst of light reaches each phototransistor simultaneously. As the light strikes,

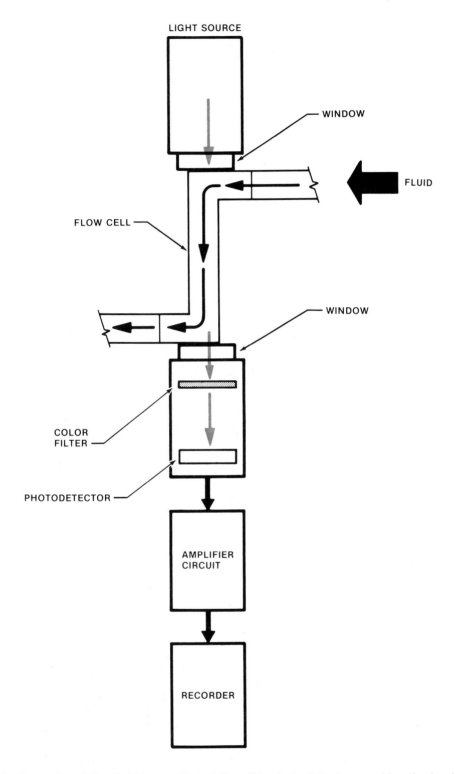

LIGHT SOURCE

WINDOW

FLUID

FLOW CELL

WINDOW

COLOR
FILTER

PHOTODETECTOR

AMPLIFIER
CIRCUIT

RECORDER

Figure 12-56. The intensity of the light beam that strikes the photodetector provides the basis for measuring the amount of light to indicate the density of the fluid.

each phototransistor conducts and discharges a capacitor into the gate of its corresponding SCR through a resistor. Since each phototransistor operates simultaneously with one another, the entire string of

SCRs turns ON simultaneously. Because there are direct connections between the light source and the input to the phototransistor, very high voltages (60 kV) can be controlled from remote locations.

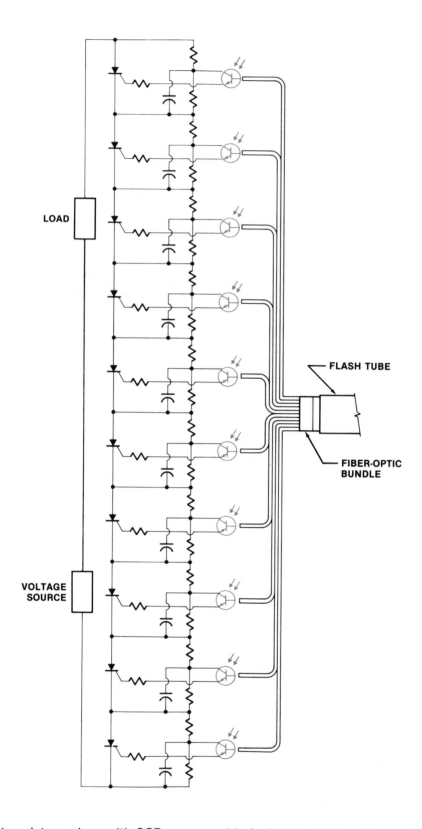

Figure 12-57. Phototransistors, along with SCRs, can provide for isolated high-voltage switching.

Chapter 12 - Review Questions

1. What is fiber optics?
2. List three advantages of fiber optics.
3. Which two areas of the spectrum are used in fiber optics?
4. How fast does light travel in free space?
5. Define refraction.
6. How is reflection accomplished in an optical fiber?
7. Describe the typical construction of an optical fiber.
8. What are typical light sources for a fiber optic system?
9. What care should be exercised in mounting an LED?
10. Define attenuation.
11. List two causes of attenuation.
12. Name the two most common types of splices.
13. How does a coupler differ from a splice?
14. What is a splitter/combiner?
15. How critical is alignment of light source and the PN junction of a PIN photodiode?
16. What is a spectral response curve?
17. What is an advantage of the phototransistor? What is a disadvantage?
18. What is an LASCR?
19. What is an optocoupler?
20. What are the three major types of optocoupler packages?

Index